Homework Book, **Foundation 2**

Delivering the Ed[illegible]ification

NEW GCSE MATHS
Edexcel Modular

Fully [illegible]ification

Brian Speed • Keith Gordon • Kevin Evans • Trevor Senior

CONTENTS

CORE

UNIT 2

UNIT 3

INTRODUCTION

Welcome to Collins New GCSE Maths for Edexcel Modular Foundation Homework Book 2. This book follows the structure of the Edexcel Modular Foundation Student Book. The first part of this book covers the Core content you need for your Unit 2 and Unit 3 exams. The second part of this book provides homework questions to cover topics in Unit 2 and Unit 3.

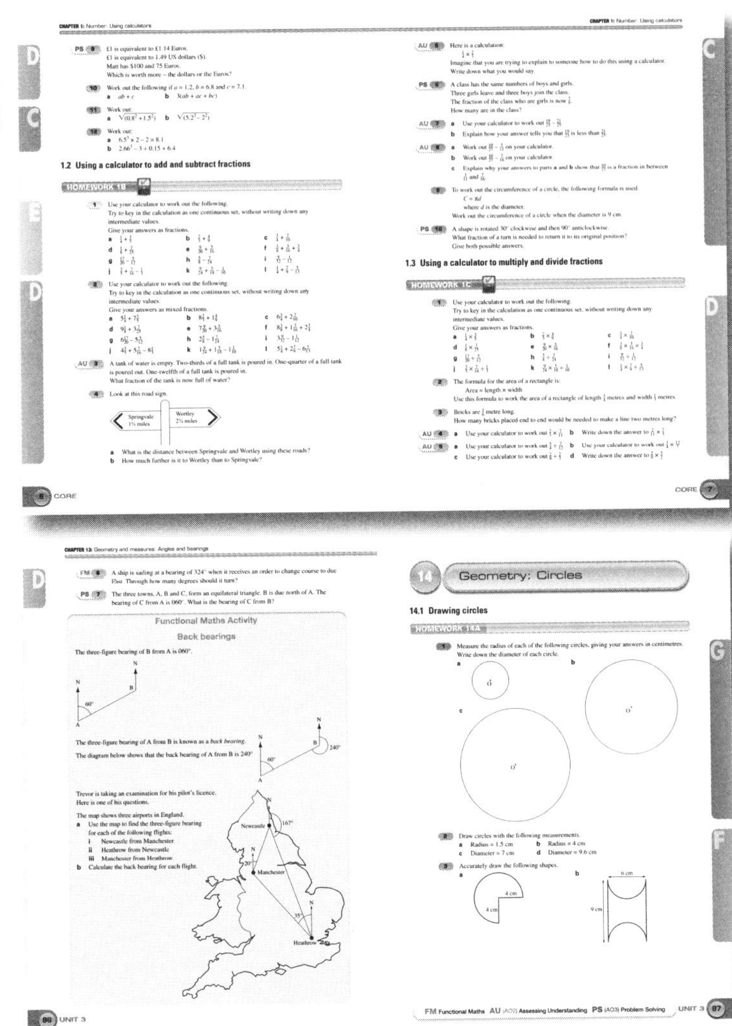

Colour-coded grades

Know what target grade you are working at and track your progress with the colour-coded grade panels at the side of the page.

Use of calculators

Questions when you could use a calculator are marked with a calculator icon. Remember in your Unit 2 exam you will not be allowed to use a calculator.

Examples

Recap on methods you need by reading through the examples before starting the homework exercises.

Functional maths

Practise functional maths skills to see how people use maths in everyday life. Look out for practice questions marked **FM**.

There are also extra functional maths and problem-solving activities at the end of every chapter to build and apply your skills.

New Assessment Objectives

Practise new parts of the curriculum (Assessment Objectives AO2 and AO3) with questions that assess your understanding marked **AU** and questions that test if you can solve problems marked **PS**. You will also practise some questions that involve several steps and where you have to choose which method to use; these also test AO2. There are also plenty of straightforward questions (AO1) that test if you can do the maths.

Student Book CD-ROM

Remind yourself of the work covered in class with the Student Book in electronic form on the CD-ROM. Insert the CD into your machine, click 'Open a PDF file' and choose the chapter you need.

Number: Using a calculator

1.1 Basic calculations and using brackets

HOMEWORK 1A

Use your calculator to work out the following questions. Try to key in the calculation in as one continuous set, without writing down any intermediate values.

1 Subtract these sets of numbers from 180.

a 90, 23 **b** 16, 57, 22 **c** 87, 36, 24

2 Subtract these sets of numbers from 360.

a 86, 21 **b** 180, 29, 97 **c** 86, 17, 17

3
a Subtract 74 from 180 and divide the answer by 2.
b Subtract 88 from 360 and divide the answer by 2.
c Subtract 2 lots of 56 from 180.
d Subtract 52 and 2 lots of 28 from 360.

4 Work out:

a $(18 - 5) \times 360 \div 24$ **b** $360 - (180 \div 3)$

5 Work out:

a $\frac{1}{2} \times (6.4 + 9.2) \times 3.6$ **b** $\frac{1}{2} \times (1.7 + 11.5) \times 7.3$

6 Work out the following and give your answers to one decimal place.

a $\pi \times 7.8$ **b** $2 \times \pi \times 6.1$ **c** $\pi \times 10.2^2$
d $\pi \times 1.9^2$

FM 7 A monthly travel ticket costs £61.60.
Karen usually spends £4.70 each day on travel.
How many days would she need to travel each month so that it would be cheaper for her to buy a monthly travel ticket?

AU 8 A teacher asked her class to work out: $\dfrac{3.1 + 5.2}{1.9 + 0.3}$

Alfie keyed in:

(3 . 1 + 5 . 2) ÷ 1 . 9 + 0 . 3 =

Becky keyed in:

3 . 1 + 5 . 2 ÷ (1 . 9 + 0 . 3) =

Chloe keyed in:

3 . 1 + 5 . 2 ÷ 1 . 9 + 0 . 3 =

Daniel keyed in:

(3 . 1 + 5 . 2) ÷ (1 . 9 + 0 . 3) =

They each rounded their answers to three decimal places.
Work out the answer that each of them got.
Who had the correct answer?

PS 9 £1 is equivalent to £1.14 Euros.
£1 is equivalent to 1.49 US dollars ($).
Matt has $100 and 75 Euros.
Which is worth more – the dollars or the Euros?

10 Work out the following if $a = 1.2$, $b = 6.8$ and $c = 7.1$.

a $ab + c$ **b** $3(ab + ac + bc)$

11 Work out:

a $\sqrt{(0.8^2 + 1.5^2)}$ **b** $\sqrt{(5.2^2 - 2^2)}$

12 Work out:

a $6.5^3 \times 2 - 2 \times 8.1$

b $2.66^3 - 3 \div 0.15 + 6.4$

1.2 Using a calculator to add and subtract fractions

HOMEWORK 1B

1 Use your calculator to work out the following.
Try to key in the calculation as one continuous set, without writing down any intermediate values.
Give your answers as fractions.

a $\frac{1}{4} + \frac{3}{5}$ **b** $\frac{2}{3} + \frac{4}{9}$ **c** $\frac{3}{4} + \frac{7}{10}$

d $\frac{1}{8} + \frac{7}{25}$ **e** $\frac{9}{20} + \frac{5}{16}$ **f** $\frac{3}{8} + \frac{3}{16} + \frac{3}{4}$

g $\frac{17}{20} - \frac{5}{12}$ **h** $\frac{5}{8} - \frac{7}{24}$ **i** $\frac{9}{32} - \frac{1}{12}$

j $\frac{3}{5} + \frac{7}{16} - \frac{1}{3}$ **k** $\frac{9}{24} + \frac{5}{18} - \frac{1}{10}$ **l** $\frac{1}{4} + \frac{7}{9} - \frac{5}{13}$

2 Use your calculator to work out the following.
Try to key in the calculation as one continuous set, without writing down any intermediate values.
Give your answers as mixed fractions.

a $5\frac{1}{4} + 7\frac{3}{5}$ **b** $8\frac{2}{3} + 1\frac{4}{9}$ **c** $6\frac{3}{4} + 2\frac{7}{10}$

d $9\frac{1}{8} + 3\frac{7}{25}$ **e** $7\frac{9}{20} + 3\frac{5}{16}$ **f** $8\frac{3}{8} + 1\frac{3}{16} + 2\frac{3}{4}$

g $6\frac{17}{20} - 5\frac{5}{12}$ **h** $2\frac{5}{8} - 1\frac{7}{24}$ **i** $3\frac{9}{32} - 1\frac{1}{12}$

j $4\frac{3}{5} + 5\frac{7}{16} - 8\frac{1}{3}$ **k** $1\frac{9}{24} + 1\frac{5}{18} - 1\frac{1}{10}$ **l** $5\frac{1}{4} + 2\frac{7}{9} - 6\frac{5}{13}$

AU 3 A tank of water is empty. Two-thirds of a full tank is poured in. One-quarter of a full tank is poured out. One-twelfth of a full tank is poured in.
What fraction of the tank is now full of water?

4 Look at this road sign.

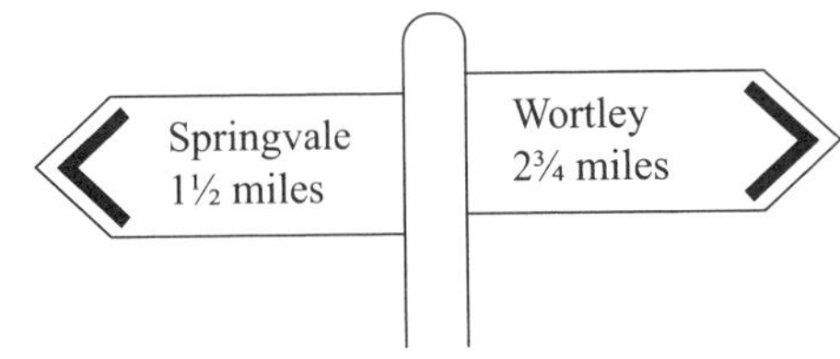

a What is the distance between Springvale and Wortley using these roads?

b How much further is it to Wortley than to Springvale?

C

AU **5** Here is a calculation:

$\frac{1}{4} \times \frac{2}{3}$

Imagine that you are trying to explain to someone how to do this using a calculator. Write down what you would say.

PS **6** A class has the same numbers of boys and girls.
Three girls leave and three boys join the class.
The fraction of the class who are girls is now $\frac{3}{8}$.
How many are in the class?

AU **7** **a** Use your calculator to work out $\frac{19}{23} - \frac{21}{25}$.

b Explain how your answer tells you that $\frac{19}{23}$ is less than $\frac{21}{25}$.

AU **8** **a** Work out $\frac{10}{27} - \frac{3}{11}$ on your calculator.

b Work out $\frac{10}{27} - \frac{7}{16}$ on your calculator.

c Explain why your answers to parts **a** and **b** show that $\frac{10}{27}$ is a fraction in between $\frac{3}{11}$ and $\frac{7}{16}$.

9 To work out the circumference of a circle, the following formula is used.

$C = \pi d$

where d is the diameter.

Work out the circumference of a circle when the diameter is 9 cm.

PS **10** A shape is rotated 30° clockwise and then 90° anticlockwise.
What fraction of a turn is needed to return it to its original position?
Give both possible answers.

1.3 Using a calculator to multiply and divide fractions

HOMEWORK 1C

D

1 Use your calculator to work out the following.
Try to key in the calculation as one continuous set, without writing down any intermediate values.
Give your answers as fractions.

a $\frac{1}{4} \times \frac{3}{5}$ **b** $\frac{2}{3} \times \frac{4}{9}$ **c** $\frac{3}{4} \times \frac{7}{10}$

d $\frac{1}{8} \times \frac{7}{25}$ **e** $\frac{9}{20} \times \frac{5}{16}$ **f** $\frac{3}{8} \times \frac{3}{16} \times \frac{3}{4}$

g $\frac{17}{20} \div \frac{5}{12}$ **h** $\frac{5}{8} \div \frac{7}{24}$ **i** $\frac{9}{32} \div \frac{1}{12}$

j $\frac{3}{5} \times \frac{7}{16} \div \frac{1}{3}$ **k** $\frac{9}{24} \times \frac{5}{18} \div \frac{1}{10}$ **l** $\frac{1}{4} \times \frac{7}{9} \div \frac{5}{13}$

2 The formula for the area of a rectangle is:

Area = length × width

Use this formula to work the area of a rectangle of length $\frac{3}{4}$ metres and width $\frac{1}{3}$ metres.

3 Bricks are $\frac{1}{6}$ metre long.
How many bricks placed end to end would be needed to make a line two metres long?

AU **4** **a** Use your calculator to work out $\frac{2}{3} \times \frac{7}{11}$ **b** Write down the answer to $\frac{2}{11} \times \frac{7}{3}$

AU **5** **a** Use your calculator to work out $\frac{3}{4} \div \frac{7}{12}$ **b** Use your calculator to work out $\frac{3}{4} \times \frac{12}{7}$

c Use your calculator to work out $\frac{2}{9} \div \frac{2}{3}$ **d** Write down the answer to $\frac{2}{9} \times \frac{3}{2}$

C

6 Use your calculator to work out the following questions. Try to key in the calculation as one continuous set, without writing down any intermediate values.
Give your answers as mixed fractions.

a $3\frac{1}{4} \times 2\frac{3}{5}$ **b** $6\frac{2}{3} \times 1\frac{4}{9}$ **c** $7\frac{3}{4} \times 2\frac{7}{10}$

d $5\frac{1}{8} \times 2\frac{7}{25}$ **e** $6\frac{9}{20} \times 4\frac{5}{16}$ **f** $1\frac{3}{8} \times 1\frac{3}{16} \times 1\frac{3}{4}$

g $4\frac{17}{20} \div 2\frac{5}{12}$ **h** $1\frac{5}{8} \div 1\frac{7}{24}$ **i** $2\frac{9}{32} \div 1\frac{1}{12}$

j $3\frac{3}{5} \times 2\frac{7}{16} \div 1\frac{1}{3}$ **k** $2\frac{9}{24} \times 3\frac{5}{18} \div 1\frac{1}{10}$ **l** $4\frac{1}{4} \times 3\frac{7}{9} \div 2\frac{5}{13}$

7 The formula for the area of a rectangle is:
Area = length × width
Use this formula to work the area of a rectangle of length $4\frac{3}{4}$ metres and width $2\frac{1}{3}$ metres.

8 The volume of a sphere is $19\frac{2}{5}$ cm^3. It is cut into four equal pieces.
Work out the volume of one of the pieces.

9 The formula for average speed is:
Average speed = Distance ÷ time taken
Work out the average speed of a car which travels $6\frac{3}{4}$ miles in a $\frac{1}{4}$ of an hour.

10 Given that 1 gallon = $4\frac{1}{2}$ litres
Grace puts 40 litres of fuel in her car.
How many gallons is this?
Give your answer to the nearest gallon.

PS FM 11 Ropes come in $12\frac{1}{2}$ metre lengths. Jack wants to cut pieces of rope that are each $\frac{3}{8}$ of a metre long.
He needs 100 pieces.
How many ropes will he need?

Functional Maths Activity

Using a calculator

The following information is written on the back of Mr Fermat's gas bill.

Reading on 19th Aug 05979
Reading on 19th Nov 06229

= 250 metric units used over 93 days

Gas units converted = 2785.52 kWh used over 93 days

First 683.00 kWh × 6.683p	£48.67
Next 2102.52 kWh × 3.292p	£69.21
Total cost of gas used	£116.08

Gas units are converted to kilowatt hours (kWh) using the following formula:

Metric units used	calorific value correction	volume	to convert to kWh	gas used in kWh
250	× 39.2236	× 1.02264	÷ 3.6	= 2785.52

Mr Fermat is having trouble understanding this and has asked for your help. Can you answer these questions for him?

1 Where does the figure of 250 units come from?

2 What does kWh stand for?

3 There are two different prices for gas. The first 683 kWh used are charged at a higher rate than the 2102.52 kWh used after that. Why are there two different prices for each kWh?

4 Can you check that the formula at the bottom has been worked out correctly: does 250 metric units convert to 2785.52 kWh of gas used?

Mr Gauss lives next door. His meter reading now is 14279 and his last meter reading was 14092.

5 Mr Gauss wants to know the cost of the gas he has used. Can you tell him?

2 Algebra: Review of algebra

2.1 Basic algebra

HOMEWORK 2A 

E

1 Simplify the following expressions.

a $3 \times 2x$ **b** $4h \times 2h$ **c** $5x + 7x$

d $2a + 5b + 6a - b$ **e** $7x + 3 - 2x - 7$

PS 2 My son is 20 years old. In five years time he will be half as old as I am. What age am I now?

3 Amy has £1.65p and Hank has £2.55. How much should Hank give to Amy so they both have the same amount?

4 Find the value of the following expressions when $a = 6$, $b = 2$ and $c = -1$.

a $5a + 1$ **b** $5b - 2a$ **c** $a^2 + c^2$

d ac **e** $2ab$ **f** $3ab - 3ac$

5 Say if the following are Expressions (E), Equations (Q) or Formulae (F).

A: $5x - 2 = 3$ B: $X + 2Y$ C: $P = 2l + 2w$

FM 6 A plumber uses the following rule to calculate his charges for jobs.

Charge = £35 plus £30 per hour.

a How much does a job that takes 3 hours cost?

b Fran pays £50 for a job. How long did it last?

c A homeowner checks the time the plumber takes to do a job. He takes 2 hours and charges £105. Has he overcharged?
Show how you decide.

D

AU 7 **a** Which of the following expressions are equivalent?

$5m \times 6n$ $3m \times 10n$ $2n \times 15m$ $m \times 30n$

b The expressions $3x$ and x^2 are the same for only one positive value of x.
What is the value?

8 Expand these expressions.

a $6(3 - 2m)$ **b** $3(x + 9)$ **c** $4n(m - 3p)$

9 Factorise the following expressions.

a $20 - 5m$ **b** $9x + 15y$ **c** $4n - pn$

PS 10 A square and a rectangle have the same perimeter.
The rectangle has one side that is three times as long as the other.
One side of the rectangle is 6 cm.
What are the two possible sides of the square?

11 Find the value of the following expressions when $x = 2.4$, $y = 0.6$ and $z = 1.5$.

a $\frac{3x + 4}{2}$ **b** $\frac{x + 2}{z}$ **c** $\frac{y}{z} + x$

FM AU **12** The formula for the electricity bill each quarter in a household is £12.25 + £0.15 per unit. A family uses 3750 units in a quarter.

a How much is their total bill?

b The family pay a direct debit of £180 per month towards their electricity costs. By how much will they be in credit or debit after the quarter?

AU **13** x and y are different positive whole numbers. Choose values for x and y so that the formula $3x + 5y$:

a evaluates to an even number

b evaluates to an odd number.

14 Kaz knows that x, y and z have the values 4, 5 and 11 but he does not know which variable has which value.

a What is the maximum that the expression $2x + y - 4z$ could be?

b What is the minimum value that the expression $4x - y + z$ could have?

15 Expand these expressions.

a $3p^2(p - 2q)$ **b** $3t^2(t^2 + 7t)$ **c** $6x(5x + 8y)$

16 Expand and simplify the following expressions.

a $3(2x - 1) + 2(x + 2)$ **b** $3(y - 1) + 4(y + 1)$

c $2(x - 2) - 3(x - 2)$ **d** $4(x + 2) + 3(x - 1)$

17 Factorise the following expressions.

a $8p^2 + 4pt$ **b** $16mp - 4m^2$

c $24a^2b + 16ab$ **d** $9a^2 - 12a + 6$

PS **18** A rectangle with sides 8 and $4x + 3$ has a smaller rectangle with sides 5 and $2x - 1$ cut from it.

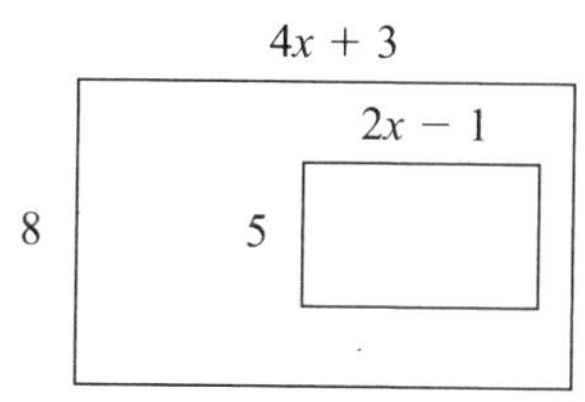

Work out the area remaining around the smaller rectangle.

2.2 Substitution using a calculator

HOMEWORK 2B

1 Find the value of $2x + 3$ when:

a $x = 2.7$ **b** $x = 3.9$ **c** $x = 8.4$

2 Find the value of $3k + 4$ when:

a $k = 6.1$ **b** $k = 12.6$ **c** $k = 18.2$

3 Find the value of A, if $A = 2t + h$ when:

a $t = 5.6, h = 3.1$ **b** $t = 9.2, h = 3.8$

4 Find the value of e, if $e = f^2 + g^2$ when:

a $e = 2.4, f = 5.1$ **b** $e = 7.2, f = 4.8$

5 Find the value of y, if $y = \sqrt{x} + n$ when:

a $x = 1.96, n = 6.7$ **b** $x = 7.29, n = 3.9$

D

FM 6 The formula $W = B + RT$ can be used to calculate a person's wage, where W is the total wage, B is the bonus, R is the hourly rate of pay, and T is the number of hours worked.

a Calculate Tom's wage if he works 35 hours for £7.80 an hour and receives a bonus of £38.50

b Calculate Maria's wage if she works $37\frac{1}{2}$ hours for £15.90 an hour and receives a bonus of £51.40

AU 7 The area of a trapezium can be calculated using the formula: $A = \frac{1}{2}h(a + b)$
Find values for a and b if $A = 21.6$ and $h = 3.6$.

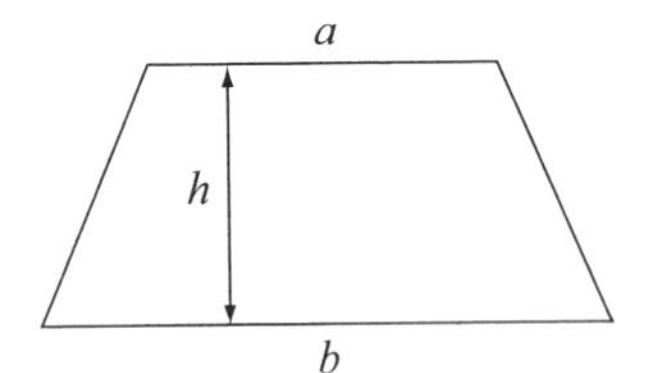

C

PS 8 The perimeter of a square $ABCD$ is 9.6 cm.
A rectangle has the same area as the square but its length is four times as long as its width.
Find the perimeter of the rectangle.

2.3 Solving linear equations

HOMEWORK 2C

Solve the following equations.

F

1 **a** $x + 9 = 5$ **b** $\frac{x}{4} = 7$

E

2 **a** $3x + 2 = 17$ **b** $\frac{x}{6} - 2 = 8$

D

3 **a** $\frac{x + 3}{z} = 10$ **b** $\frac{x - 5}{3} = 2$

4 **a** $4(x + 3) = 8$ **b** $4(3x - 5) = 10$

5 **a** $9x + 7 = 3x + 19$ **b** $5x - 2 = 16 - 4x$

C

6 **a** $4(x + 9) = x + 3$ **b** $4(x - 2) - 3(x - 1) = 2(x + 2)$

Problem-solving Activity

Eating out

Six people pay the following for their meals:

Ann:	Burger, Chips, Beans	£3.30
Bashir:	Sausage, Chips and Beans	£3.00
Carol:	Burger, Roast Potatoes and Peas	£3.30
Derek:	Burger, Chips and Peas	£3.40
Emir:	Sausage and Roast Potatoes	£2.30
Farook:	Burger and Peas	£2.50

Use this information to work out how much each item costs.

Geometry and measures: Scales

3.1 Scale drawings

HOMEWORK 3A

1 The grid below shows the floor plan of a kitchen. The scale is 1 cm to 30 cm.

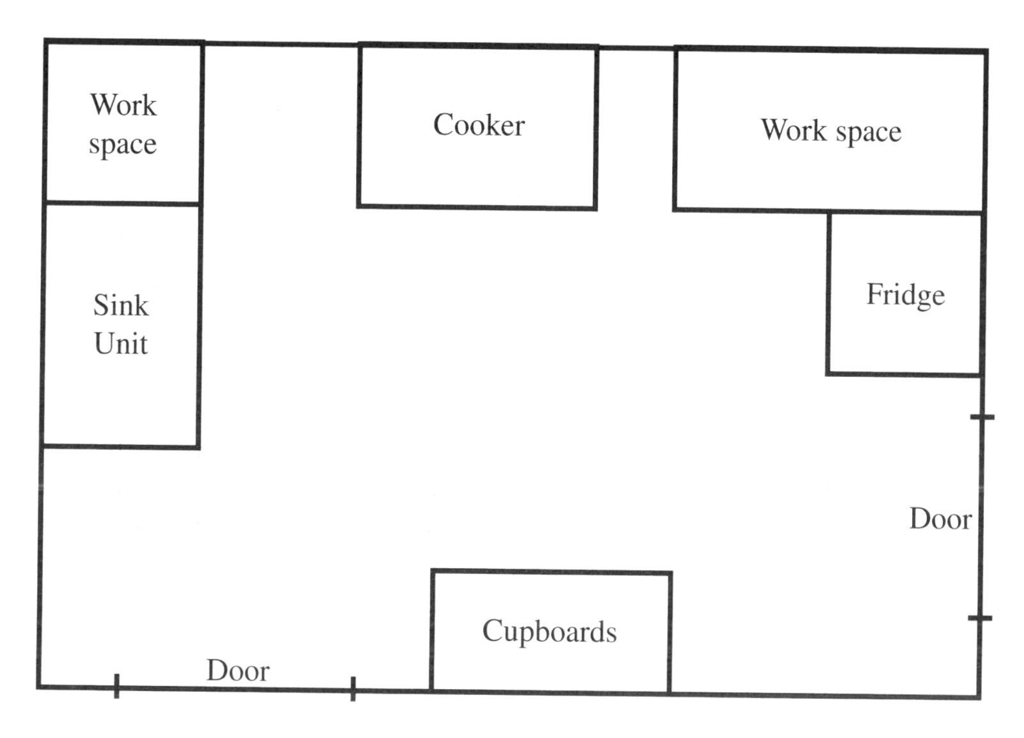

a State the actual dimensions of:

i the sink unit

ii the cooker

iii the fridge

iv the cupboards.

b Calculate the actual total area of the work space.

2 On the right is a sketch of a ladder leaning against a wall.
The bottom of the ladder is 1 m away from the wall and it reaches 4 m up the wall.

a Make a scale drawing to show the position of the ladder.
Use a scale of 4 cm to 1 m.

b Use your scale drawing to work out the actual length of the ladder.

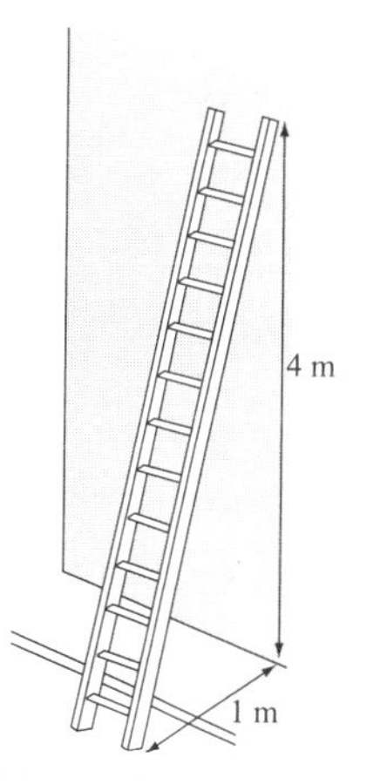

F

3 The map below is drawn to a scale of 1 cm to 2 km.

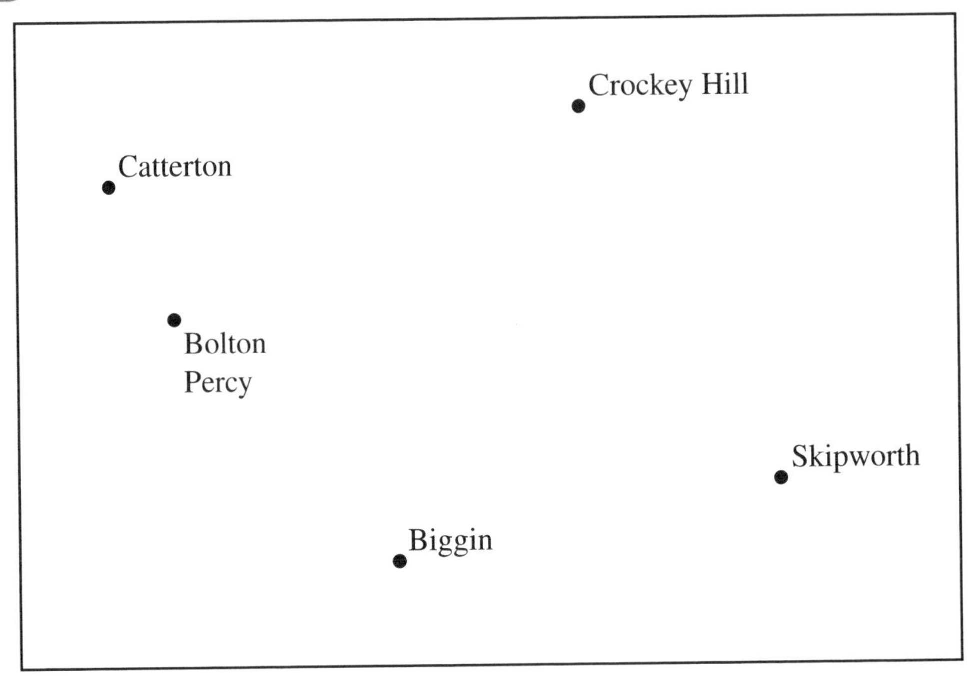

Find the distances between:

a Biggin and Skipworth

b Bolton Percy and Crockey Hill

c Skipworth and Catterton

d Crockey Hill and Biggin

e Catterton and Bolton Percy.

FM 4 A farmer draws a sketch for one of his fields. This is his diagram.

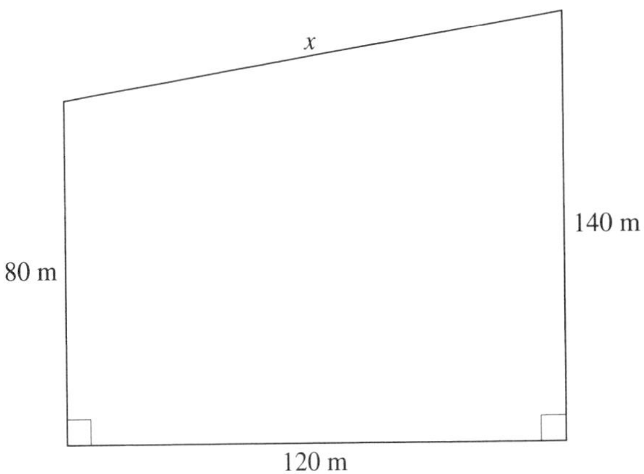

a Make a scale drawing of the field. Use the scale 1 cm represents 20 m.

b The farmer wants to build a wall along the side marked x on the diagram. Each metre length of wall uses 60 bricks.
Use your diagram to work out the number of bricks the farmer will need.

5 The map below shows the position of four fells in the Lake District. The map is drawn to a scale of 1 : 150 000.

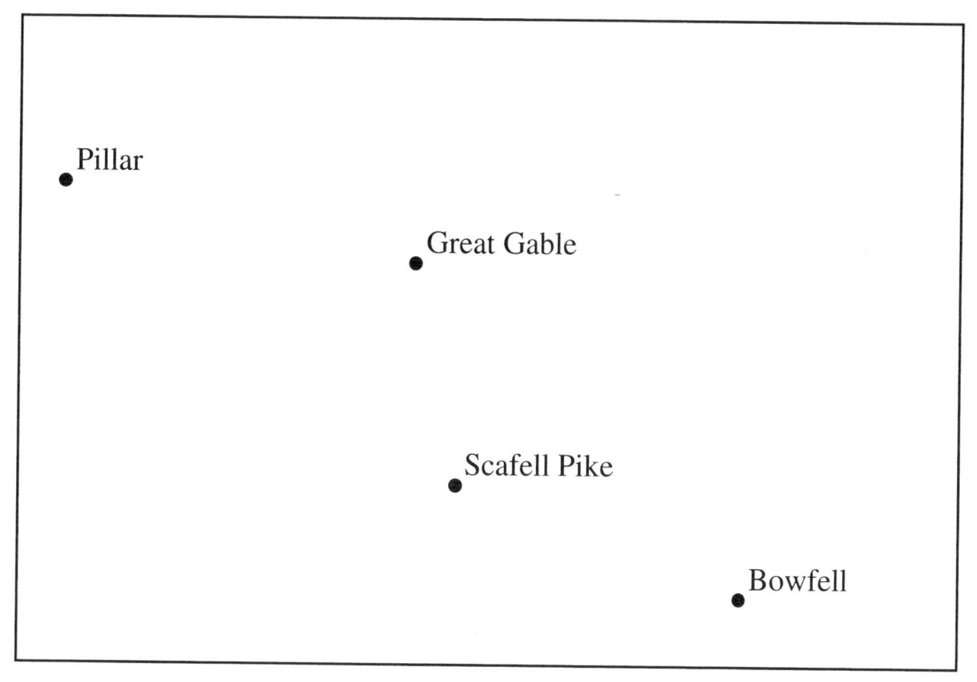

State the following distances to the nearest kilometre:

a Scafell Pike to Great Gable

b Scafell Pike to Pillar

c Great Gable to Pillar

d Pillar to Bowfell

e Bowfell to Great Gable

6 Here is a scale drawing of a ferry crossing a river from port A to port B. The width of the river is 400 m.

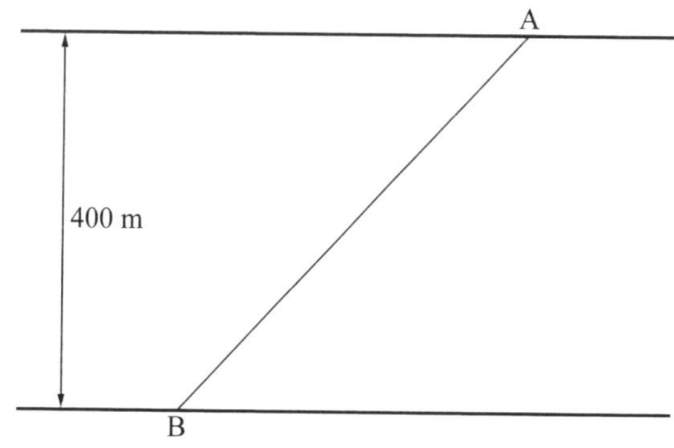

a Which of the following is the correct scale for the drawing?

i 1 : 1000 **ii** 1 : 10 000 **iii** 1 : 40 000 **iv** 1 : 100 000

b What is the actual distance from port A to port B?

3.2 Nets

HOMEWORK 3B

1 Four nets are shown below. Copy the nets that would make a cube.

a

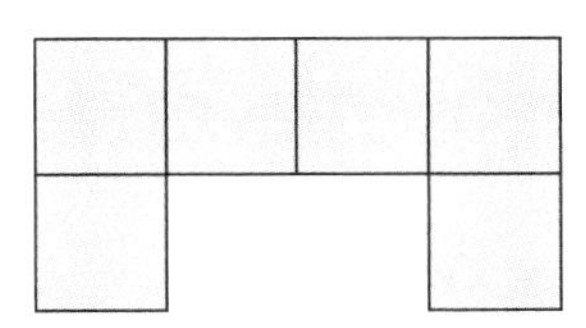

b

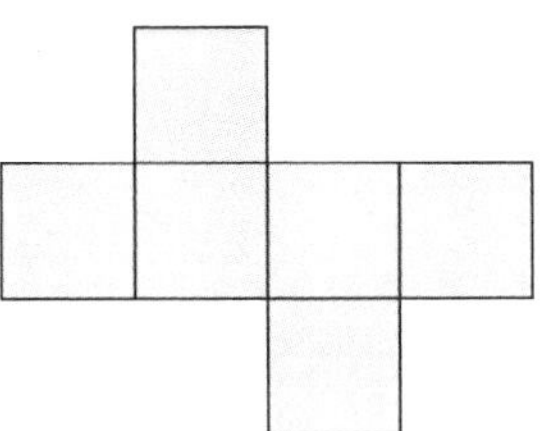

c

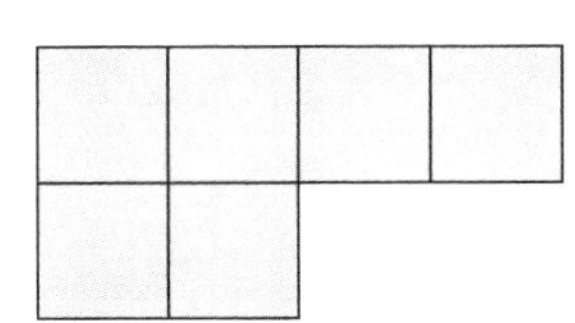

d

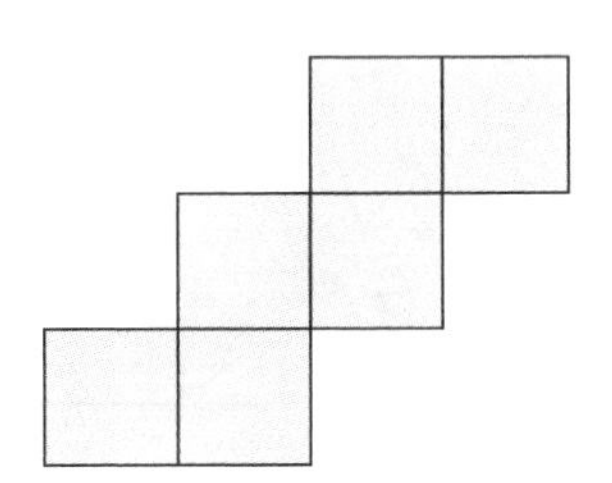

2 Draw, on squared paper, an accurate net for each of these cuboids.

a

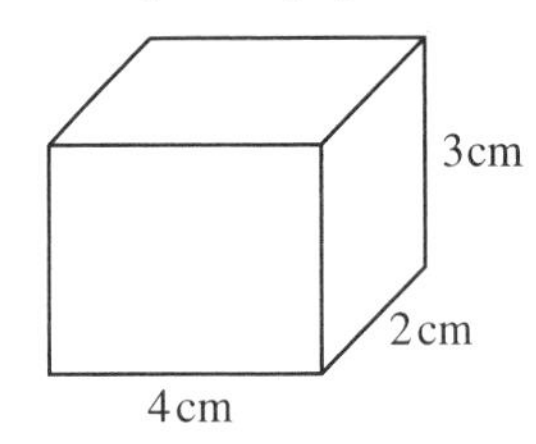

b

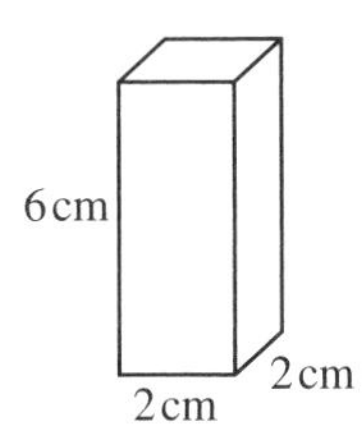

3 Draw, on squared paper, an accurate net for this triangular prism.

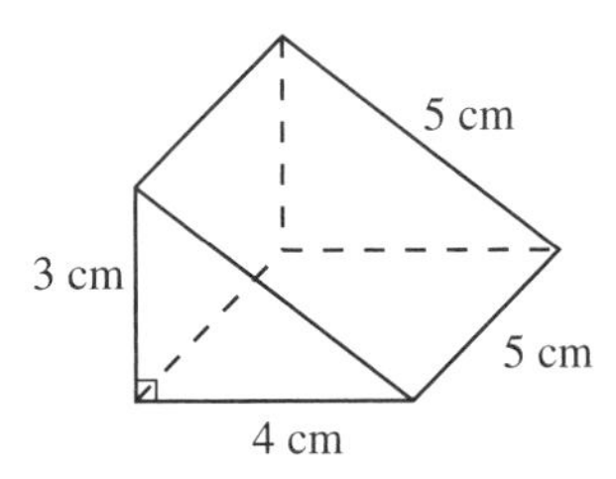

4 The diagram shows a sketch of a square-based pyramid.

a Write down how many of each of the following the pyramid has:

i vertices

ii edges

iii faces.

b Draw a sketch for the net of the pyramid.

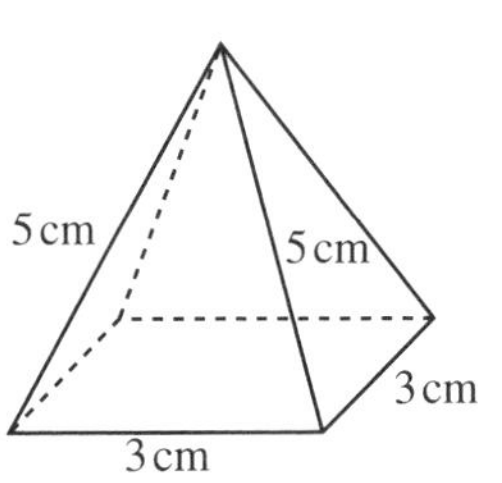

FM 5 Paul is making this dice out of card.

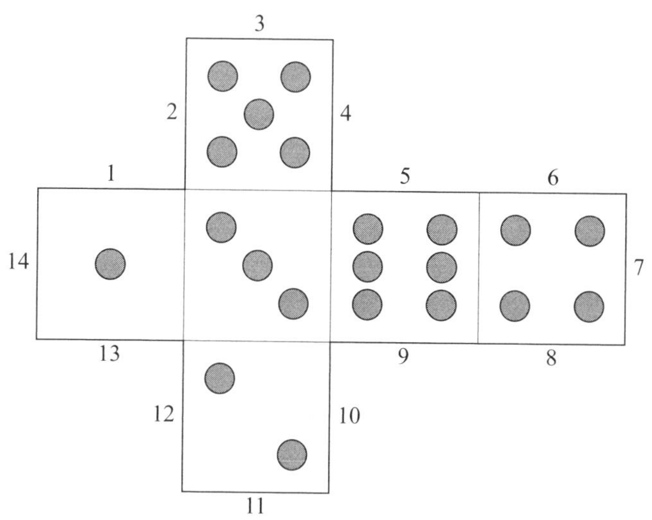

Before he cuts it out, he needs to know which edges join together so that he can put on tabs. For example, edge 1 joins with edge 2.
List the other pairs of edges that join together.

PS 6 This is a diagram of a regular tetrahedron.

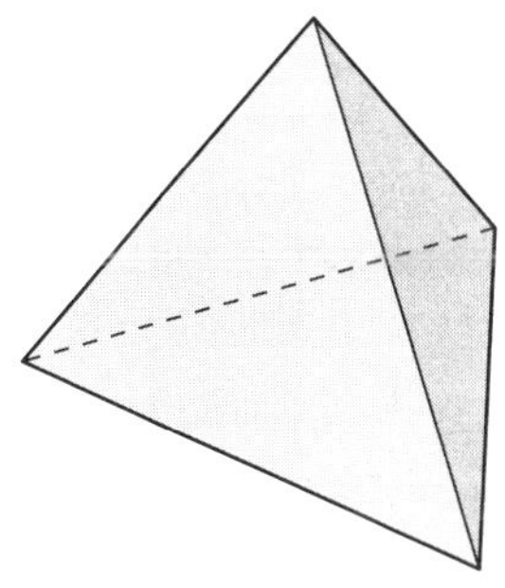

It is made up of four identical equilateral triangles.
A tetrahedron has two nets. Can you draw them both?

AU 7 Zoë has five shapes:

Two isosceles triangles: 5 cm, 5 cm, 3 cm

One rectangle 3 × 10 cm: 3 cm, 10 cm

Two rectangles 5 × 10 cm: 5 cm, 10 cm

Draw a sketch to show how she can put the five shapes together to make a net of a triangular prism.

3.3 Using an isometric grid

HOMEWORK 3C

E

1 Draw accurately each of these cuboids on an isometric grid.

a

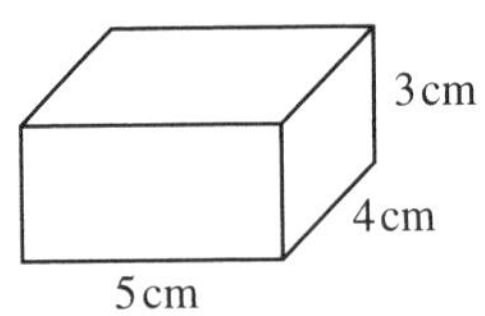

b

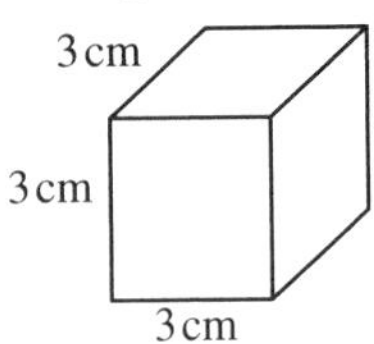

D

2 Draw accurately each of these 3D shapes on an isometric grid.

a

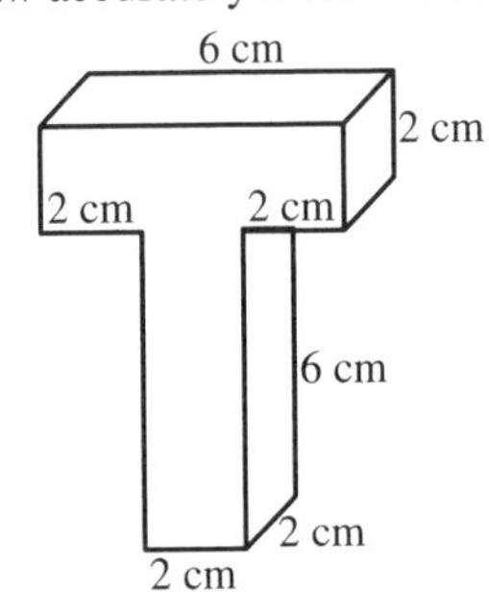

b

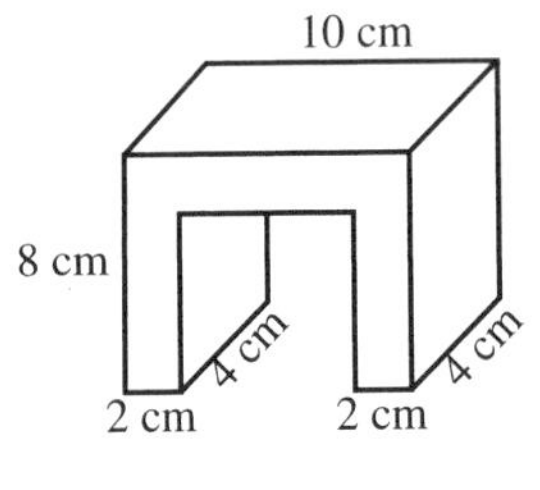

3 For each of the following 3D shapes, draw on squared paper:

i the plan **ii** the front elevation **iii** the side elevation.

a

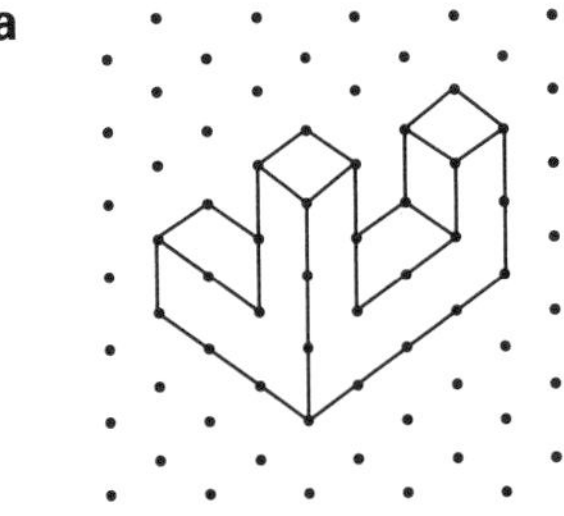

b

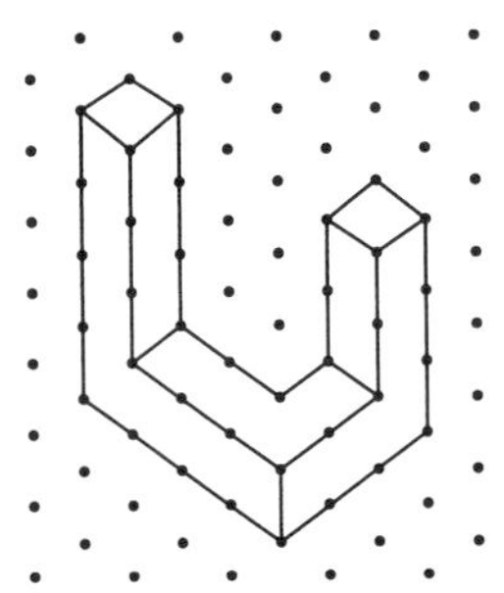

PS 4 Here are three views of a 3D shape.

Plan

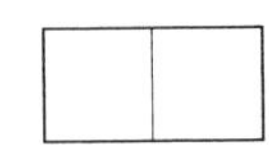

Front elevation

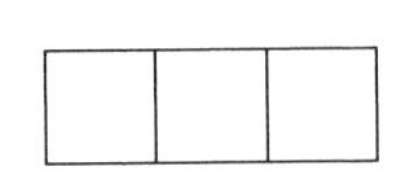

Side elevation

Draw the 3D shape on an isometric grid.

AU 5 This solid shape is made from cubes.

Plan view

Front elevation

Side elevation

Here are some diagrams of the shape.

A B

C

D F

E

a Which is the plan view?

b Which is the front elevation?

c Which is the side elevation?

Functional Maths Activity

Logo design

A firm called Trader Limited wants to design a logo for their company. An artist designs the following logo, which is drawn on isometric paper.

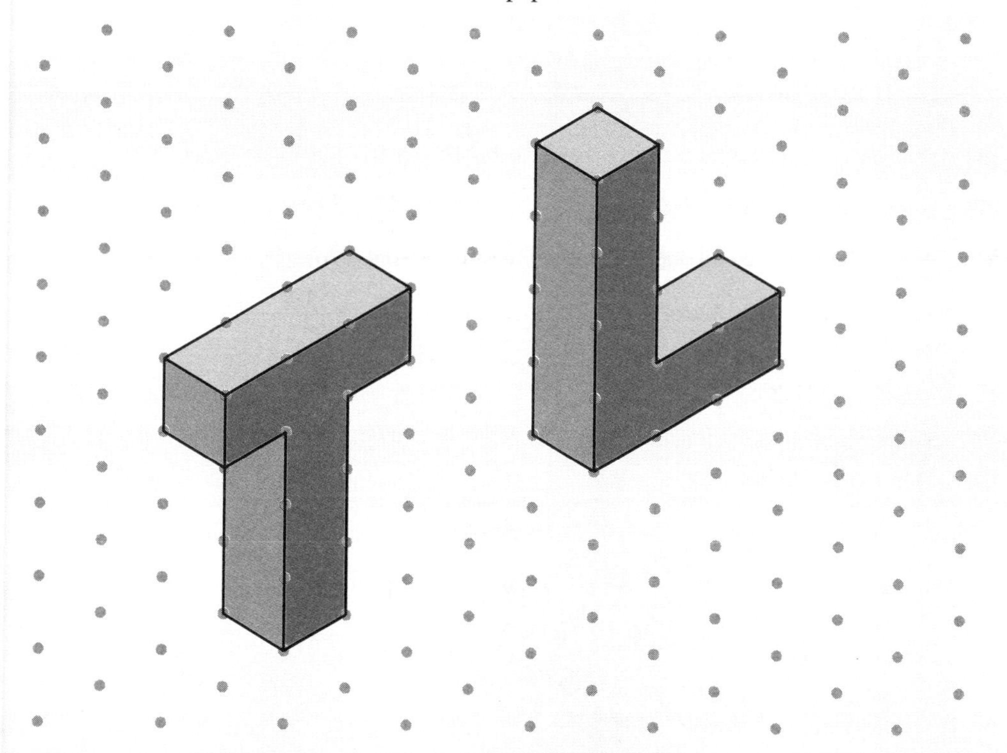

On isometric paper, design a logo for a company called 'Fisher Home Incorporated'.

4 Geometry: Perimeter and area

4.1 Perimeter

HOMEWORK 4A

G

1 Find the perimeter of each of the following shapes. Draw them on centimetre-squared paper first if it helps you.

a

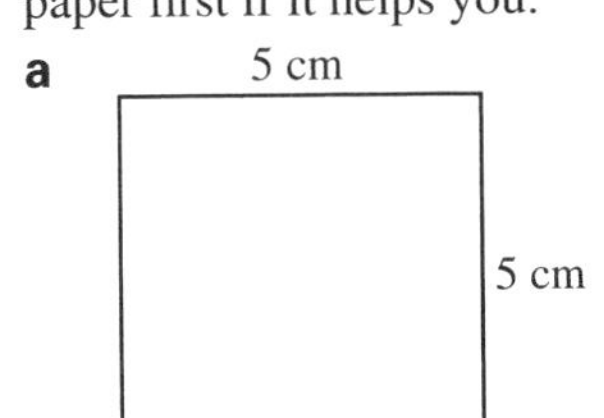

b

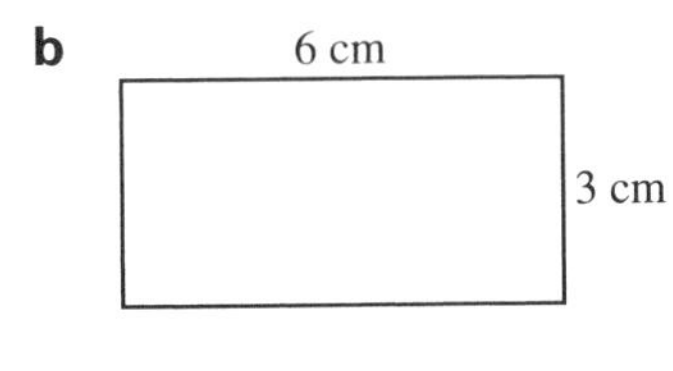

c

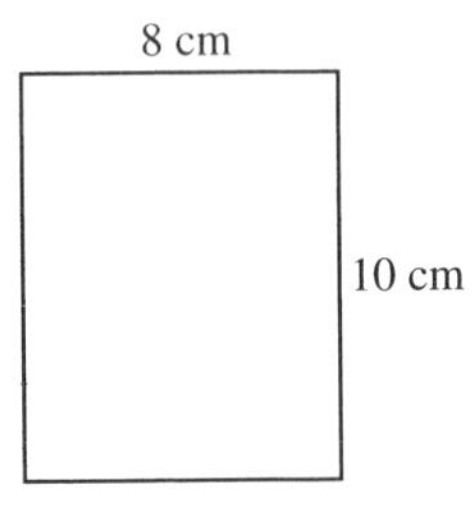

d

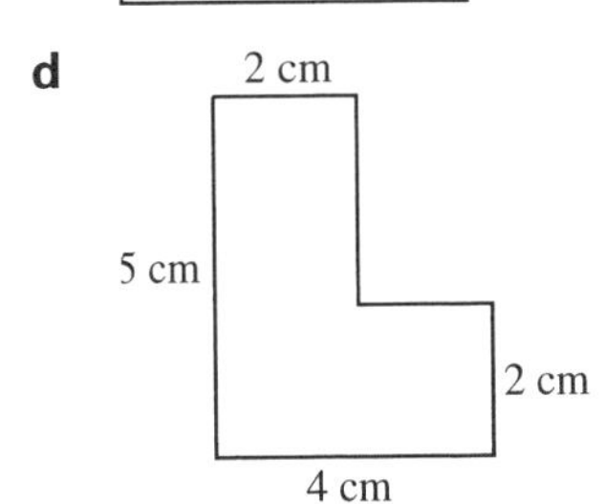

e

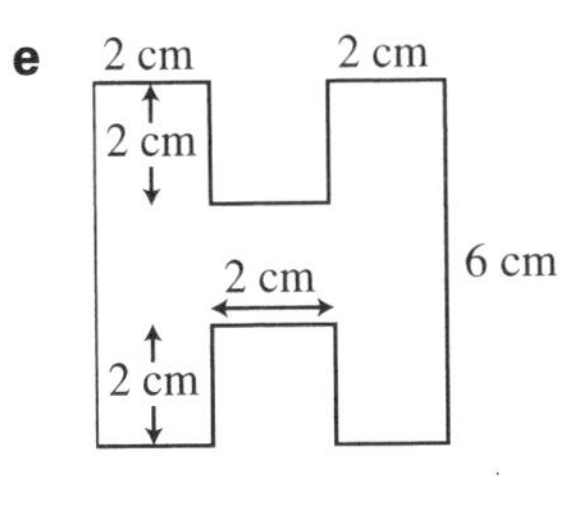

f

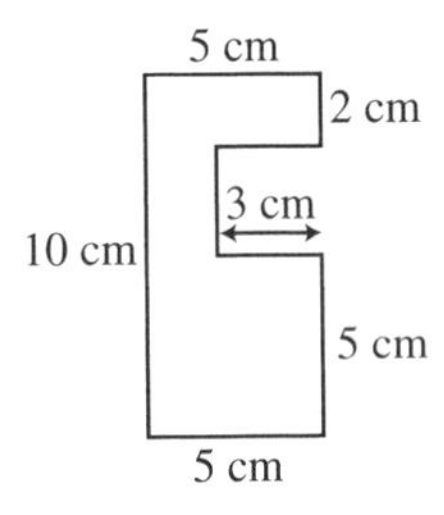

F

2 Draw as many different rectangles as possible that have a perimeter of 14 cm.

PS 3 Is it possible to draw a rectangle with a perimeter of 9 cm? Explain your answer.

AU 4 Which shape is the odd one out? Give a reason for your answer.

5 cm, 3 cm, A

7 cm, 1 cm, B

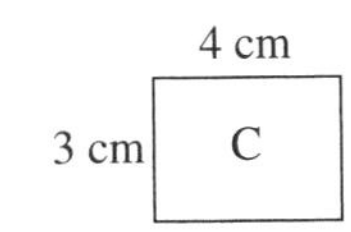

FM 5 Simon wants to put a fence around three sides of a lawn.
How much fencing does he need?

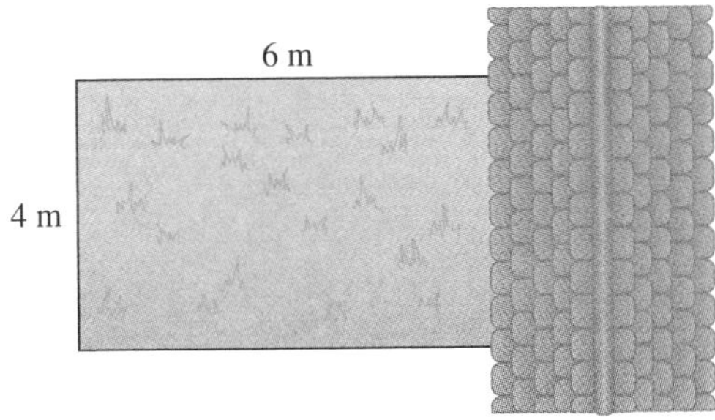

4.2 Area of an irregular shape

HOMEWORK 4B

G

1 By counting squares, find the area of each of these shapes, giving answers in cm^2.

a

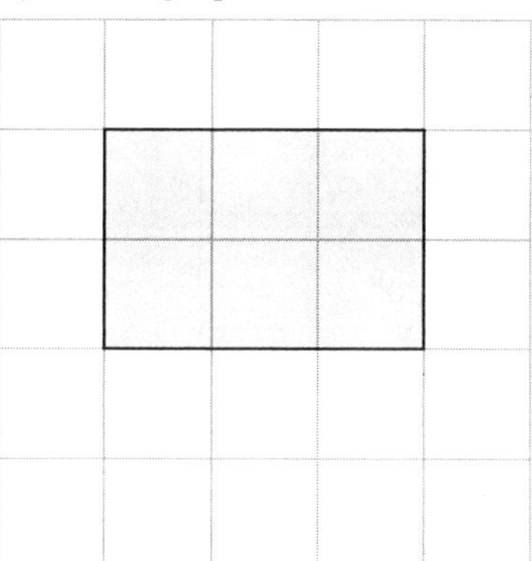

b

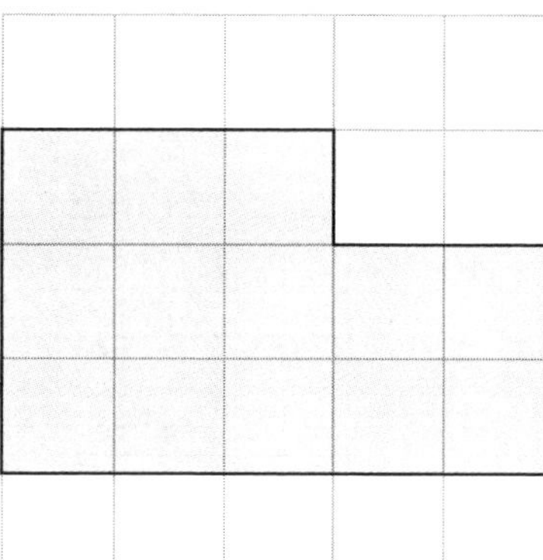

c

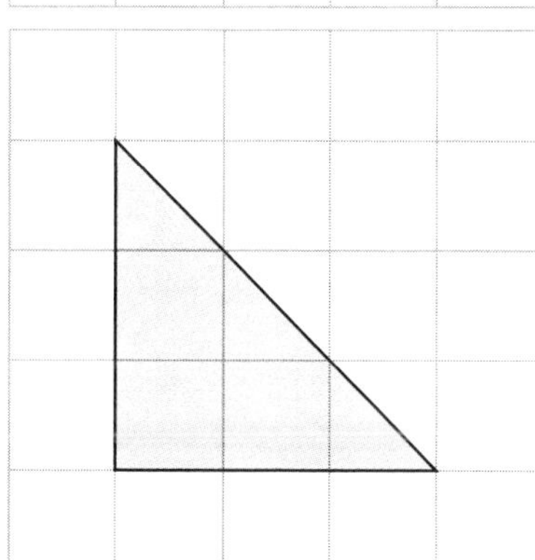

d

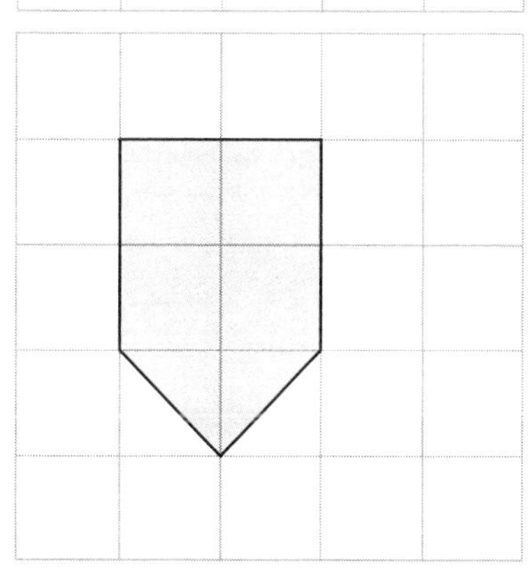

2 By counting squares, estimate the area of each of these shapes, giving answers in cm^2.

a

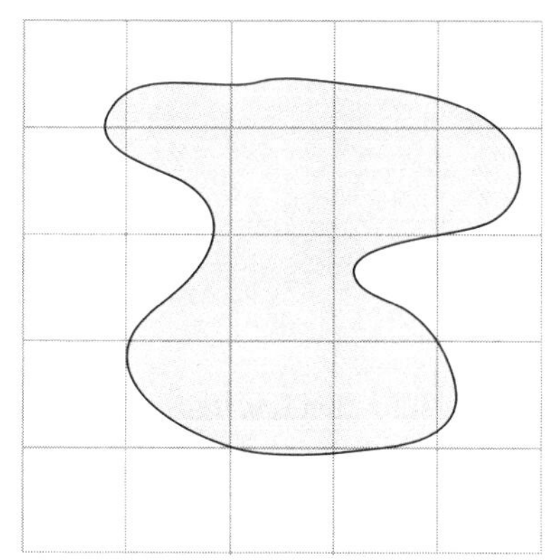

b

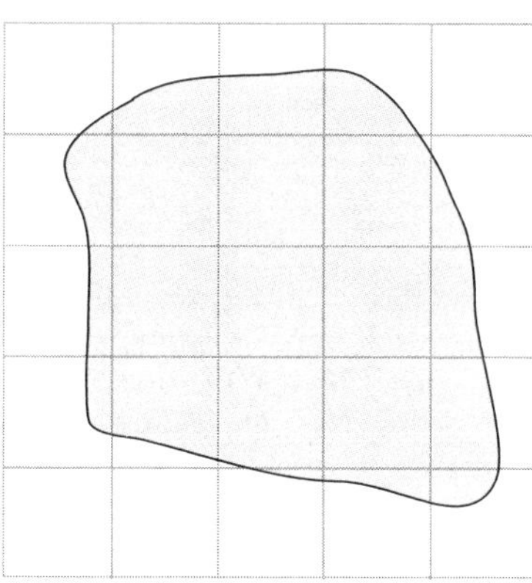

c

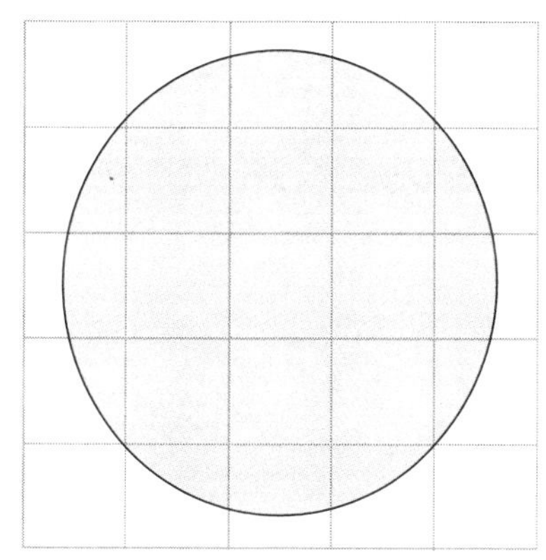

d

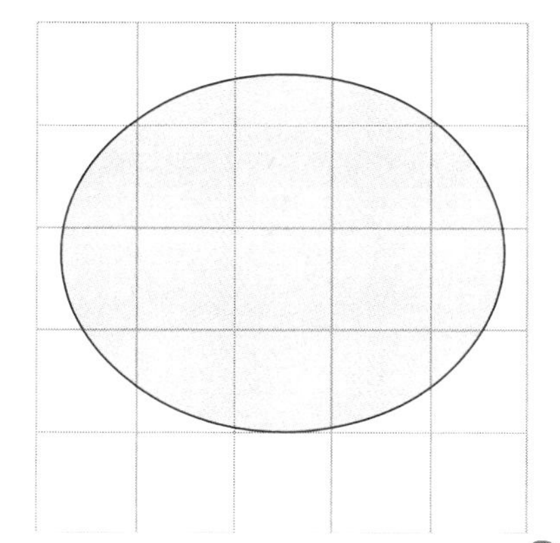

G

FM 3 Mr Bluegum, a forester, needs to find an estimate for the area of a forest.
On the map below, the forest is shown in green.

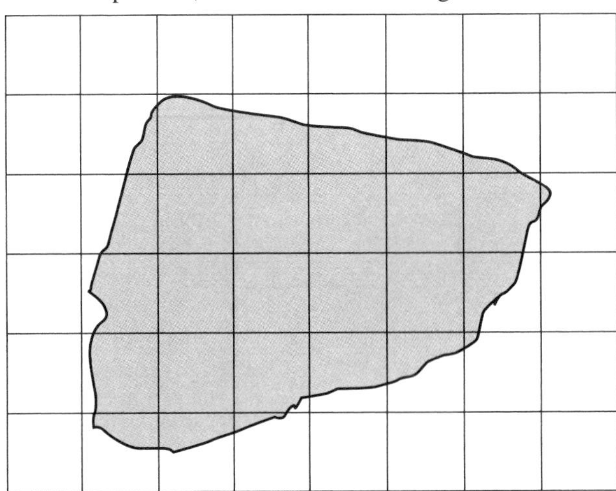

Each square on the map represents 1 km^2.
Find an estimate for the area of the forest for Mr Bluegum.

PS 4 This shape is drawn on a centimetre-squared grid.

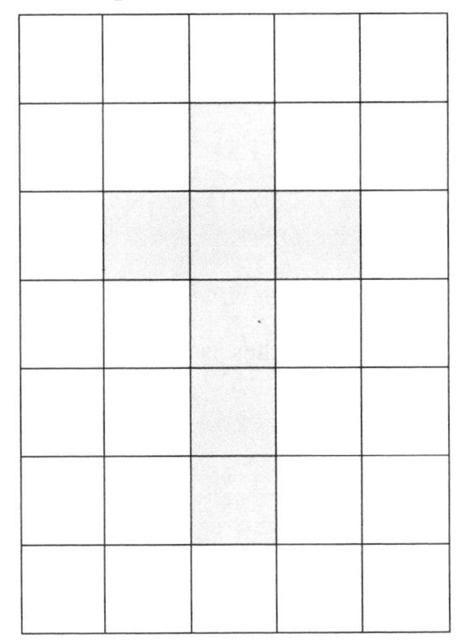

a Write down the area of the shape.

b On centimetre-squared paper, draw a square that has the same perimeter as the shape.

AU **5** George says that he can find an estimate for the area of a circle by first finding the area of a square around the circle and then finding the area of a square inside the circle. The answer is the value between these two numbers.

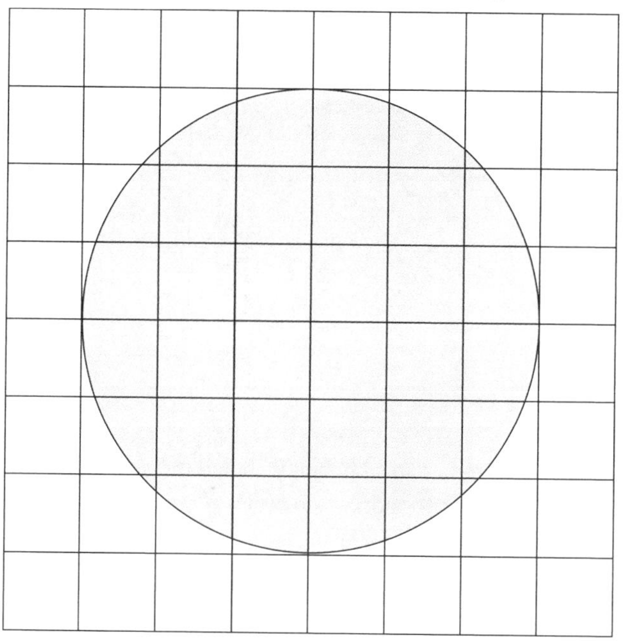

Show how George finds an estimate for the area of this circle.

4.3 Area of a rectangle

HOMEWORK 4C

1 Calculate the area and the perimeter of each rectangle below.

a

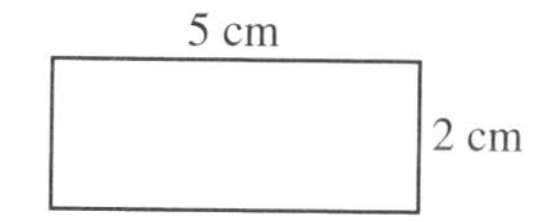

b

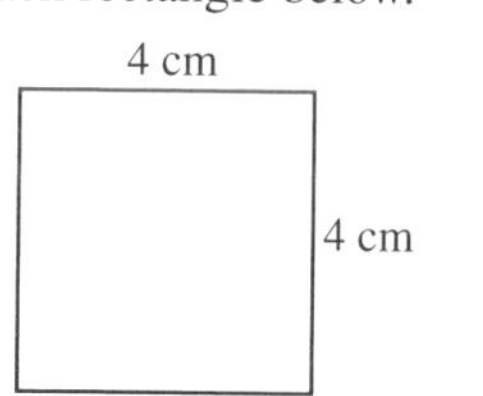

c

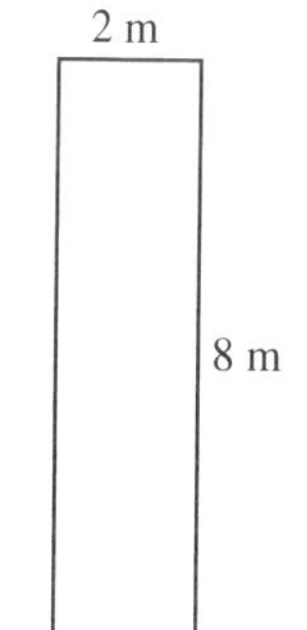

d 12 mm, 3 mm

e 20 m, 10 m

2 Copy and complete the following table for rectangles **a** to **e**.

	Length	Width	Perimeter	Area
a	4 cm	2 cm		
b	7 cm	4 cm		
c	6 cm		22 cm	
d		3 cm		15 cm^2
e			30 cm	50 cm^2

3 A square has a perimeter of 24 cm.
What is its area?

D

4 Copy and complete the following.

a **i** $1 \text{ cm}^2 = \text{..........} \text{ mm}^2$ **ii** $3 \text{ cm}^2 = \text{..........} \text{ mm}^2$ **iii** $12 \text{ cm}^2 = \text{..........} \text{ mm}^2$

b **i** $1 \text{ m}^2 = \text{..........} \text{ cm}^2$ **ii** $4 \text{ m}^2 = \text{..........} \text{ cm}^2$ **iii** $10 \text{ m}^2 = \text{..........} \text{ cm}^2$

PS **5** This shape is made from four rectangles that are all the same size.
Work out the area of one of the rectangles.

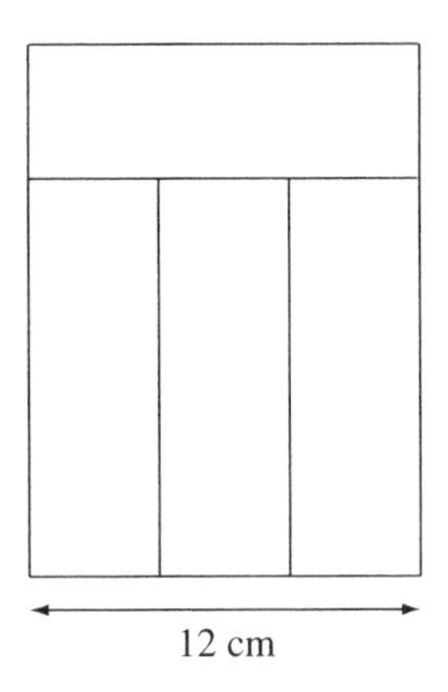

FM **6** The diagrams show the size of Lin's kitchen wall and the size of the square tile she wants to use to tile the wall. They are not drawn to scale.

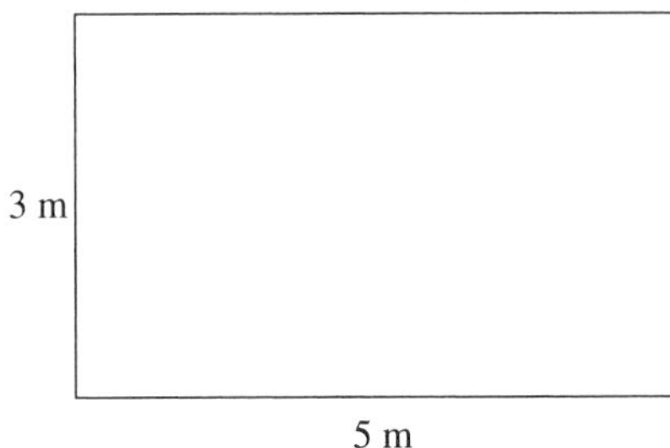

What is the minimum number of tiles Lin will need to cover the wall?

HINTS AND TIPS

Remember to change the measurements of the wall into centimetres first.

4.4 Area of a compound shape

HOMEWORK 4D

D

1 Calculate the area of each shape below.

a

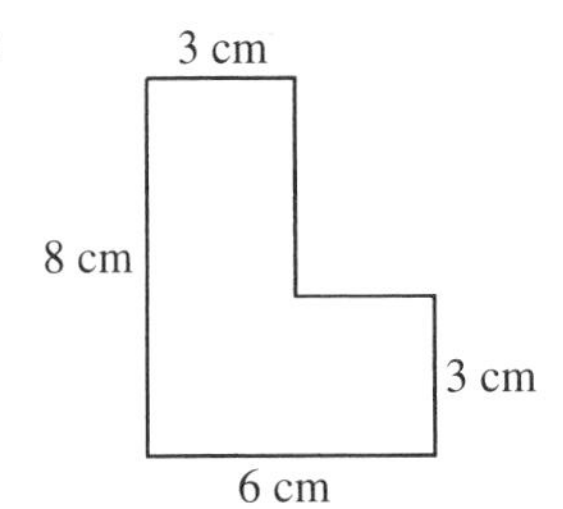

b

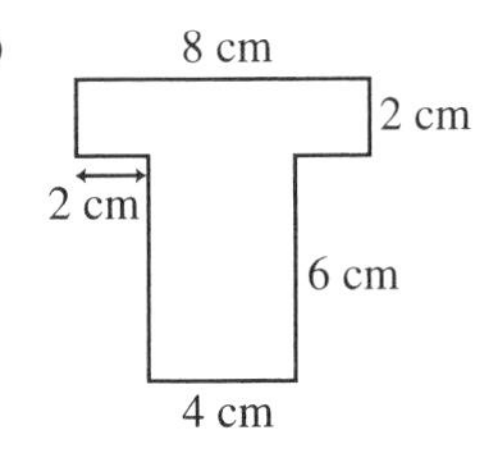

c

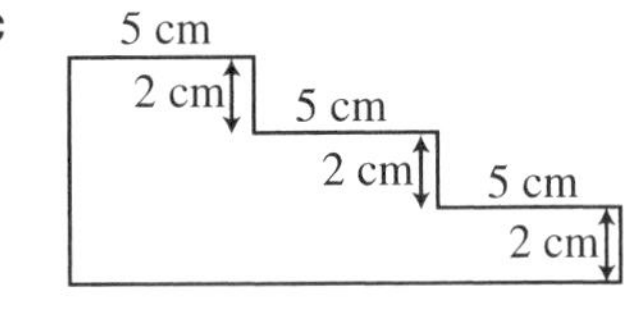

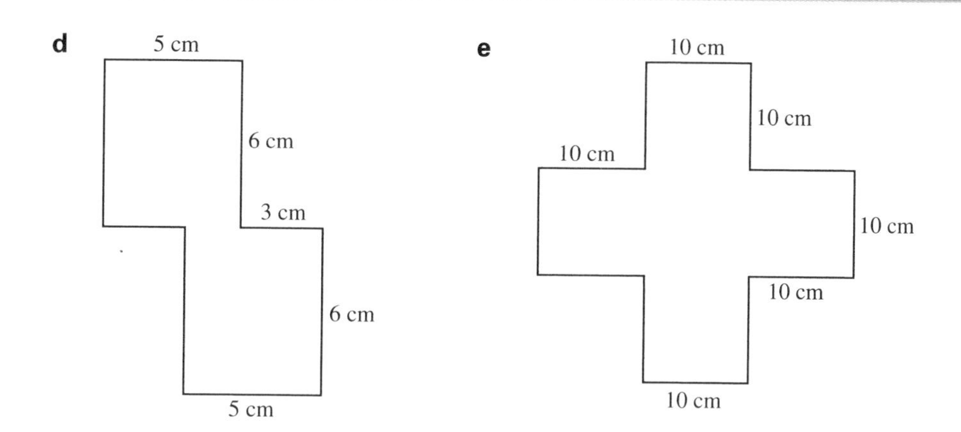

FM 2 Mr Jackson is fixing Formica® onto a worktop in his kitchen. Formica® comes in rolls 5 metres long and 0.5 metres wide.

This is a sketch of the worktop.

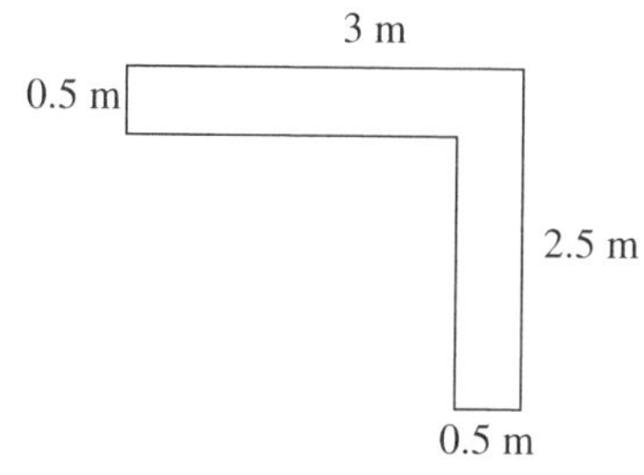

a Work out the area of the worktop.

b Does he have enough Formica® in one roll to cover his worktop?

AU 3 Rachael says that the area of this shape is 64 cm^2.

Is she correct? Give a reason for your answer.

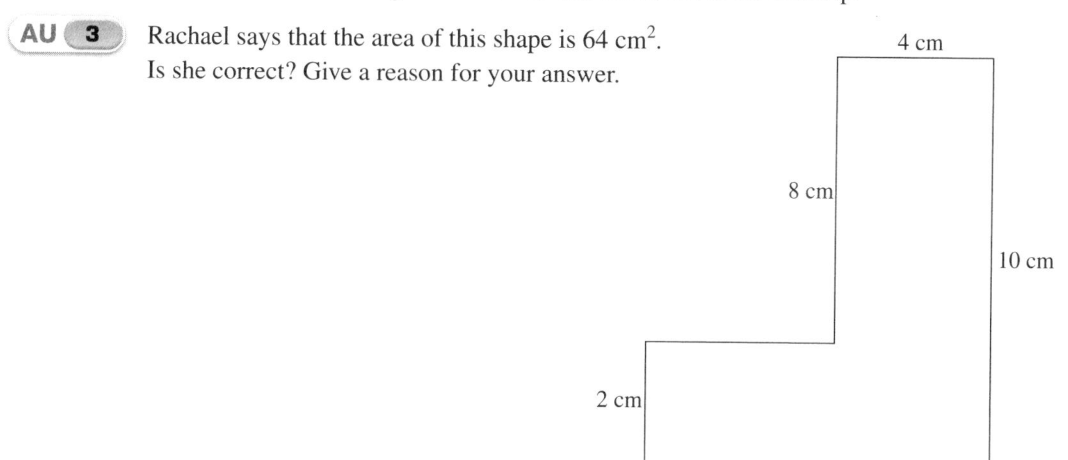

PS 4 This L-shape is made from two rectangles that are the same size.

It has an area of 48 cm^2.

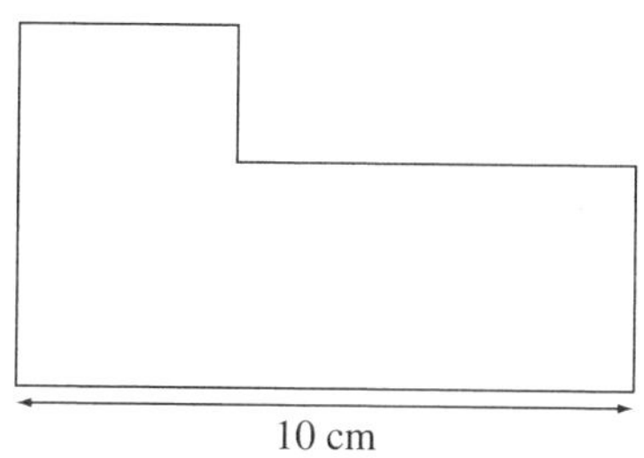

Find the length and width of each rectangle.

4.5 Area of a triangle

HOMEWORK 4E

Example Find the area of this triangle

$$\text{Area} = \tfrac{1}{2} \times 7 \times 4$$
$$= \tfrac{1}{2} \times 28 = 14 \text{ cm}^2$$

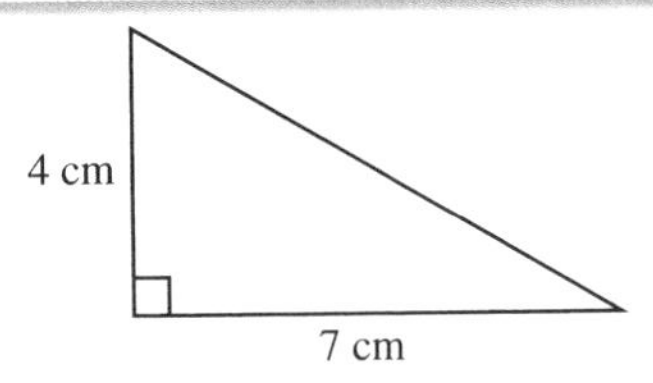

D

1 Write down the perimeter and area of each triangle.

a

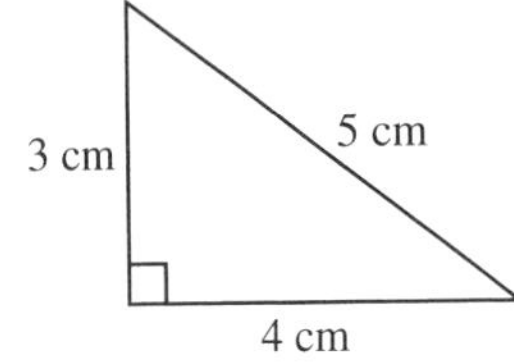

b

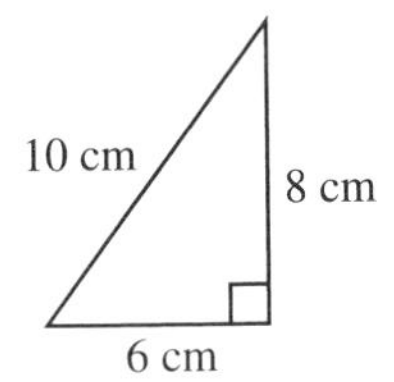

c

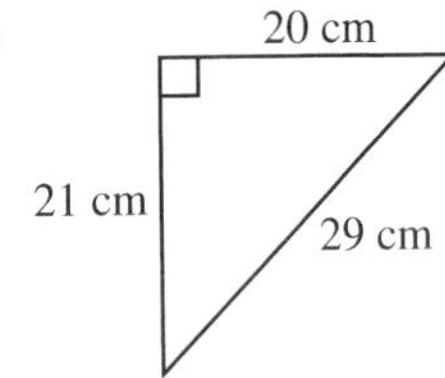

2 Work out the area of each of these compound shapes, made from rectangles and right-angled triangles.

a

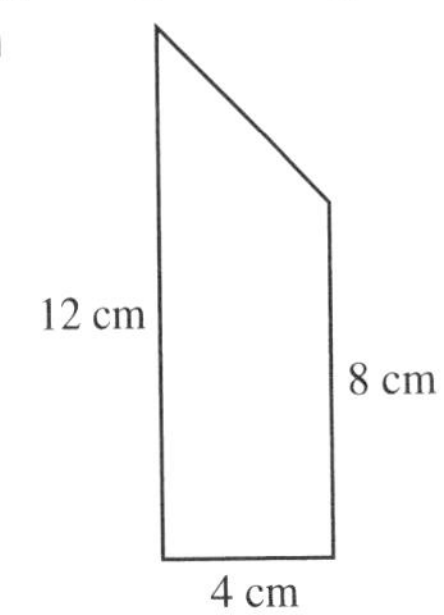

b

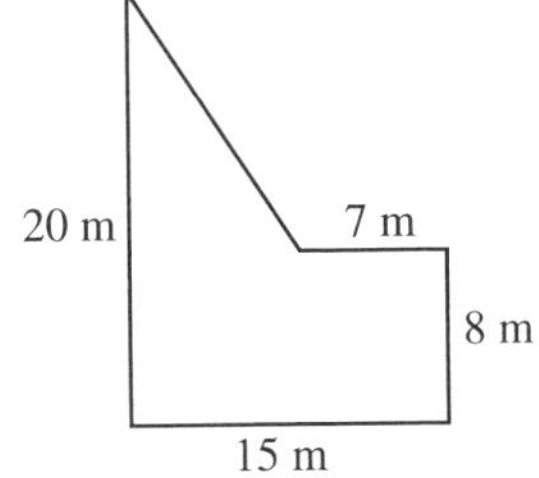

c

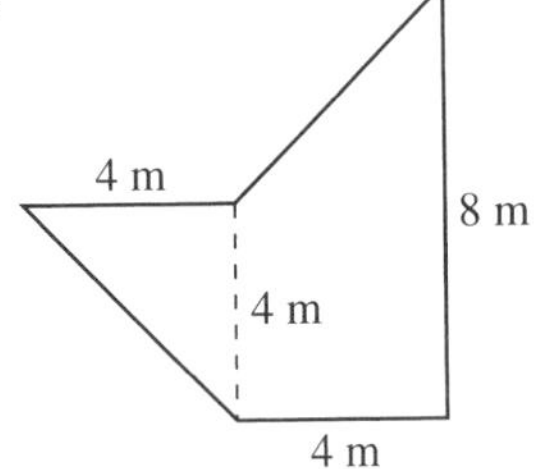

3 Find the area of the wood on this blackboard 90° set square.

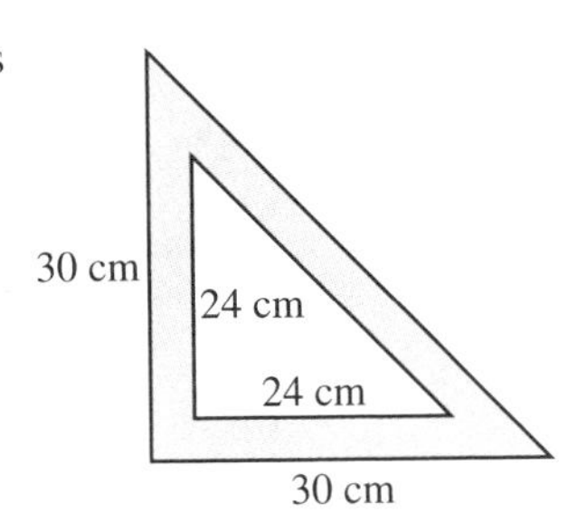

4 Which of these three triangles has the smallest area?

a

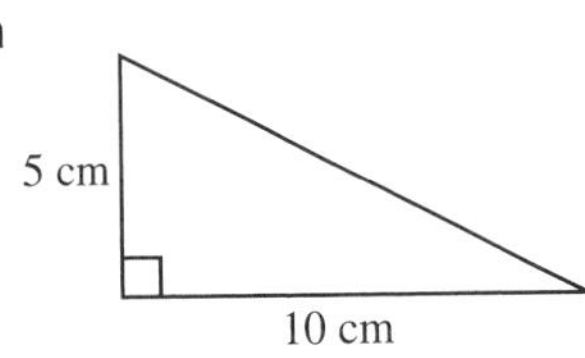

b

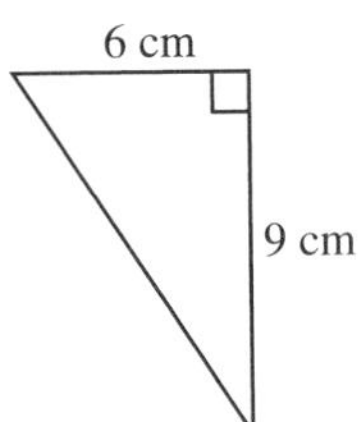

c

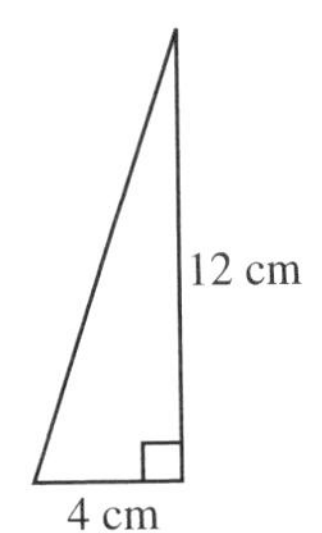

D

AU 5 Jen and Jack are comparing their answers to this question.

Work out the area of this right-angled triangle.

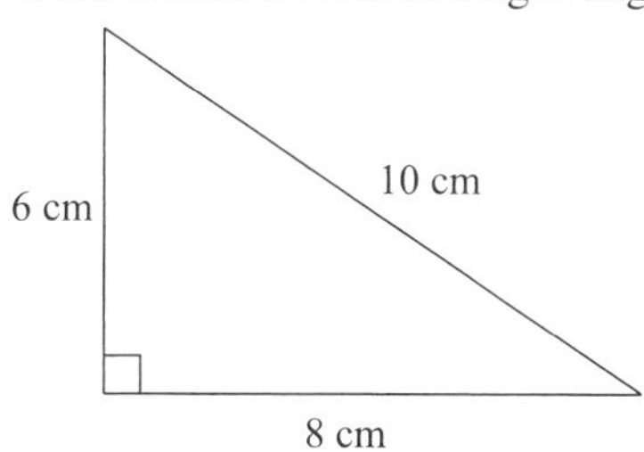

Jen's answer

$$\begin{aligned} A &= \tfrac{1}{2} \times 8 \times 6 \\ &= 4 \times 6 \\ &= 24 \text{ cm}^2 \end{aligned}$$

Jack's answer

$$\begin{aligned} A &= \tfrac{1}{2} \times 8 \times 10 \\ &= 4 \times 10 \\ &= 40 \text{ cm}^2 \end{aligned}$$

Who is correct?

Give a reason for your answer.

PS 6 Work out the area of this rhombus.

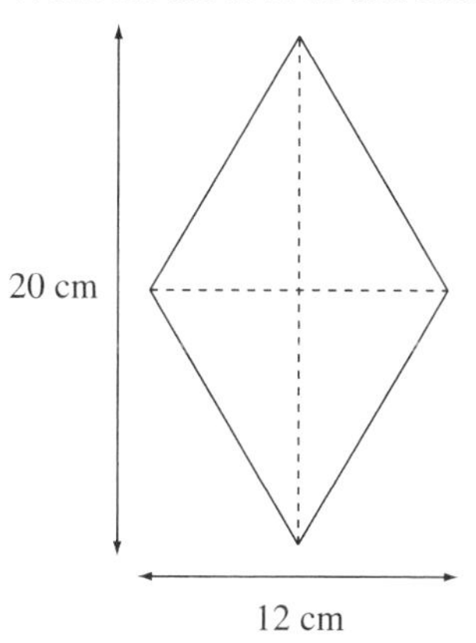

HINTS AND TIPS

The diagonals of the rhombus intersect at right angles.

HOMEWORK 4F

Example Find the area of this triangle.

$$\text{Area} = \tfrac{1}{2} \times 9 \times 4$$
$$= \tfrac{1}{2} \times 36 = 18 \text{ cm}^2$$

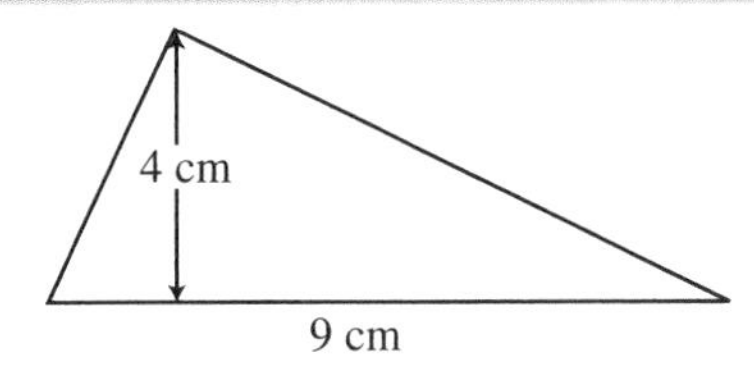

D

1 Calculate the area of each of these triangles.

a

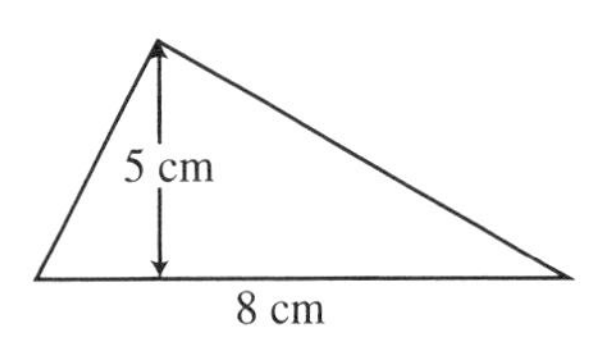

b

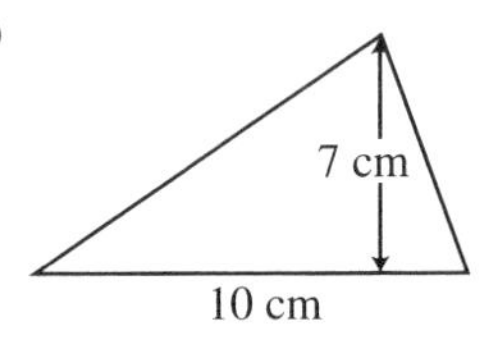

c

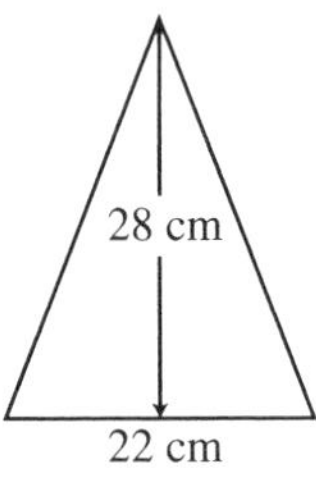

d

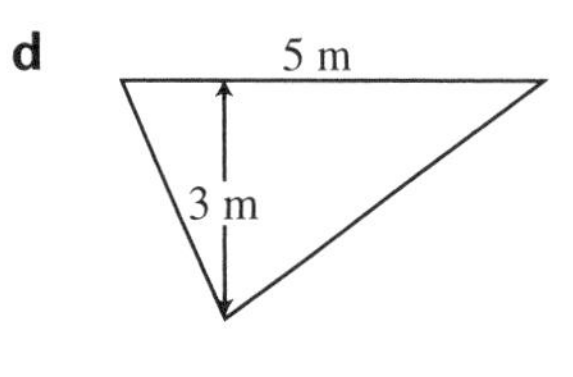

e

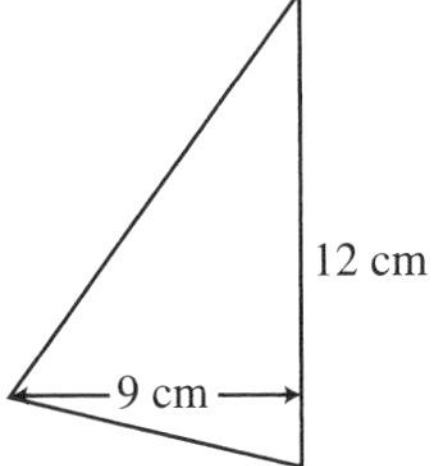

f

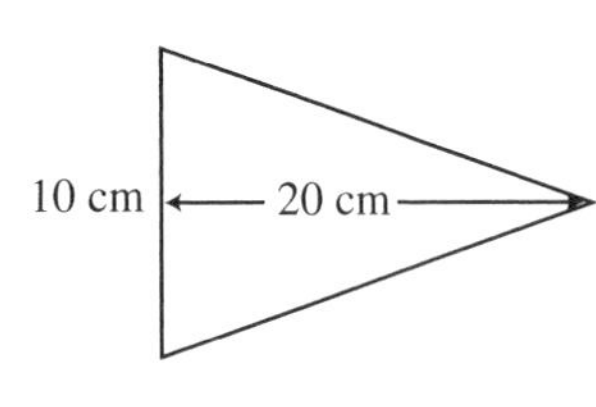

2 Copy and complete the following table for triangles **a** to **e**.

	Base	Vertical height	Area
a	6 cm	8 cm	
b	10 cm	7 cm	
c	5 cm	5 cm	
d	4 cm		12 cm^2
e		20 cm	50 cm^2

3 Find the area of each of the shaded shapes.

a

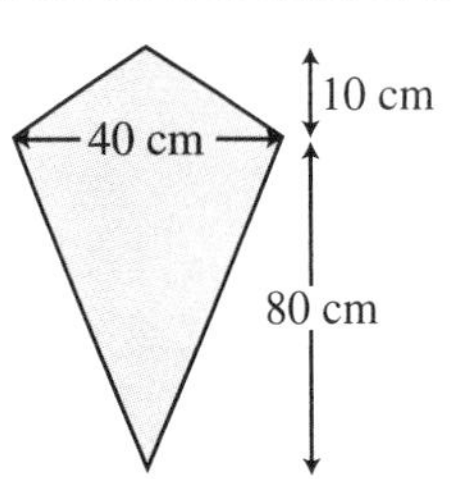

b

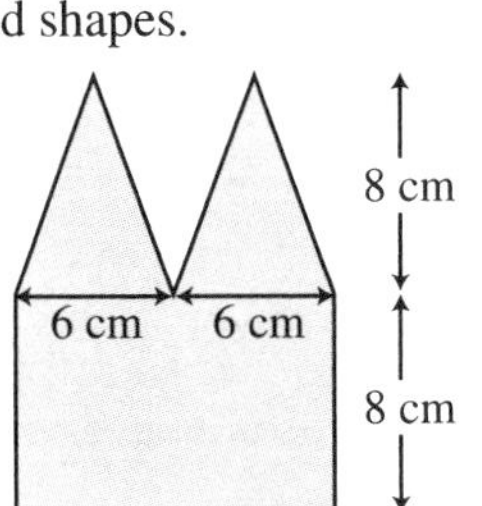

c

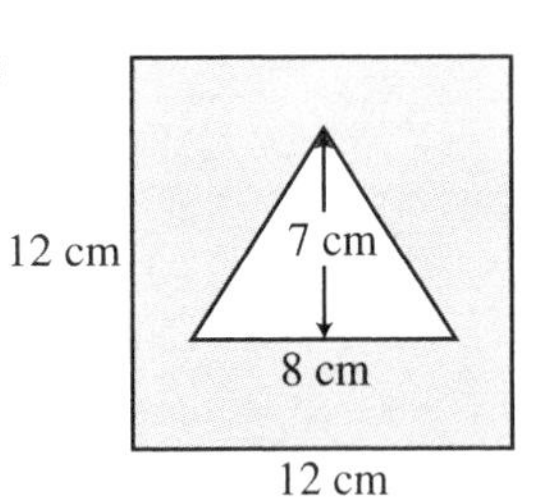

4 Draw diagrams to show two different-sized triangles that have the same area of 40 cm^2.

PS 5 The rectangle and triangle below have the same area.

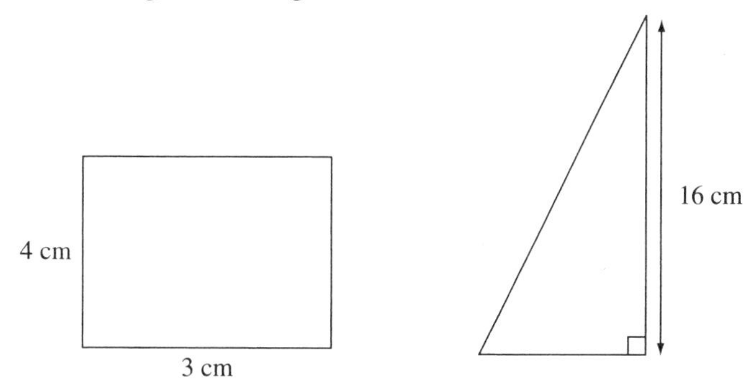

Work out the length of the base of the triangle.

AU 6 What is the same and what is different about these two triangles?

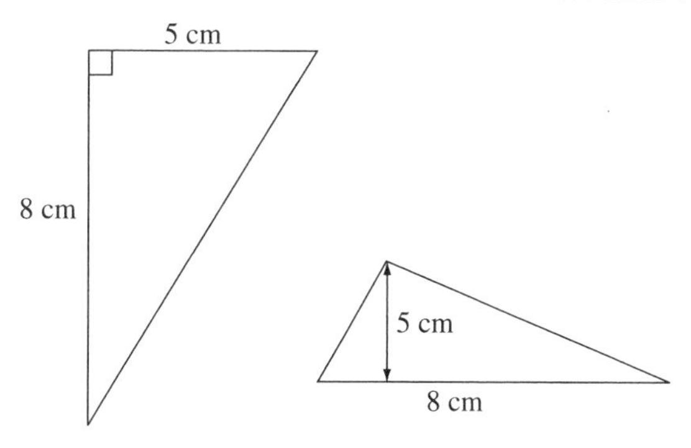

FM 7 Mary is making a mosaic from coloured tiles.

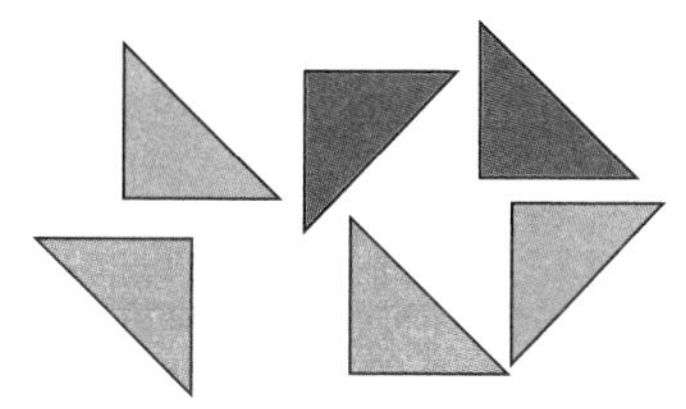

Each triangle has the following measurements:

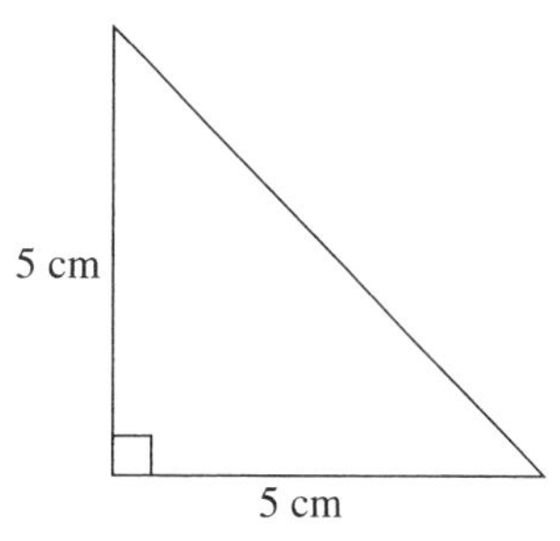

How many tiles does Mary need to cover this rectangular board completely without leaving any gaps?

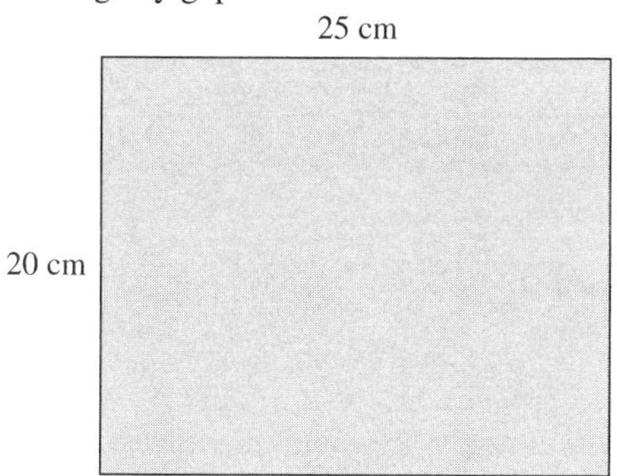

D

C

4.6 Area of a parallelogram

HOMEWORK 4G

Example Find the area of this parallelogram.

$$\text{Area} = 8 \times 6$$
$$= 48\ \text{cm}^2$$

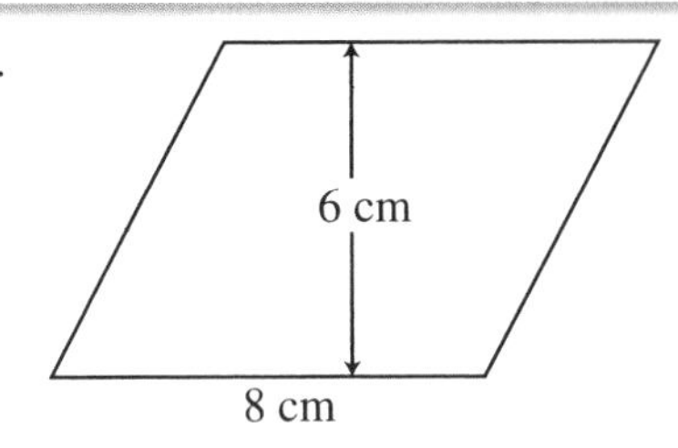

D

1 Calculate the area of each parallelogram below.

a

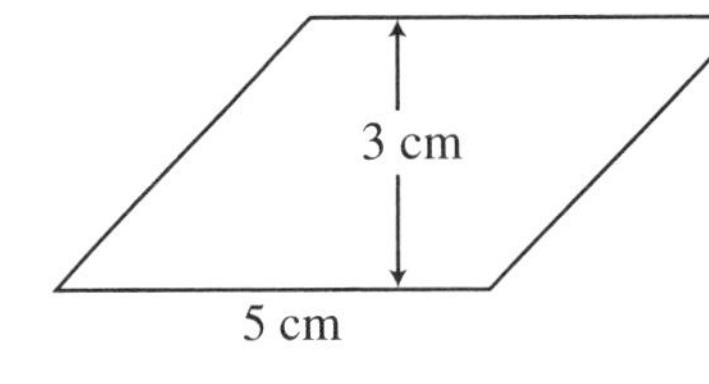

b

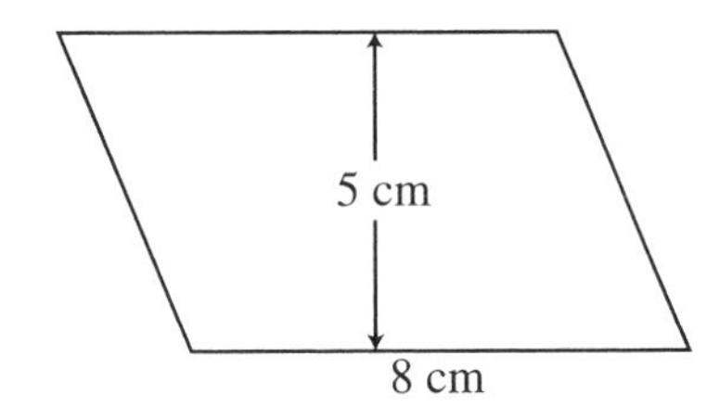

c

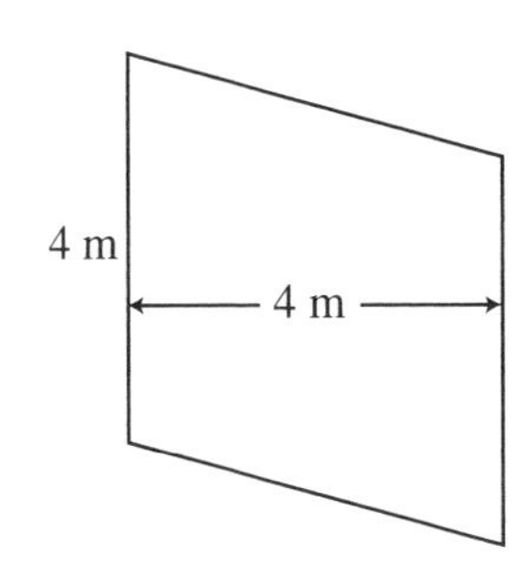

d

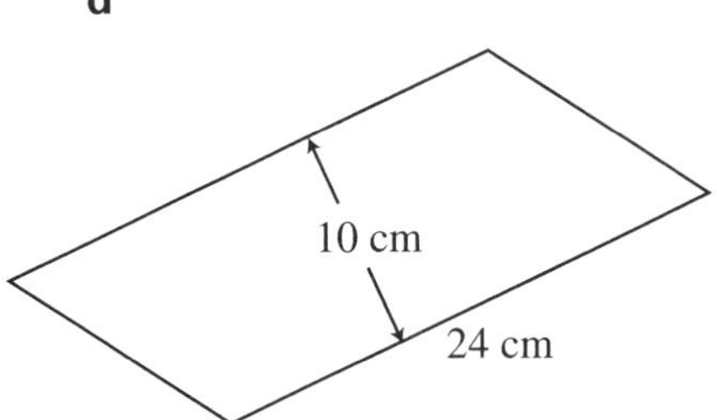

2 Find the area of the shaded section.

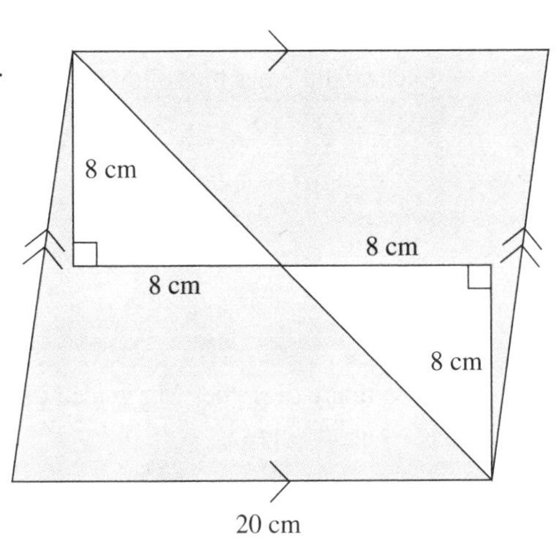

D

AU 3 Which two shapes have the same area? Show your working.

a

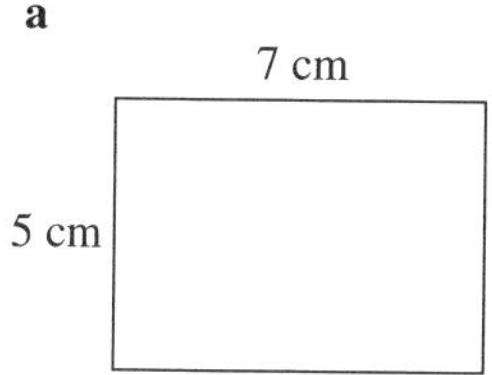

b

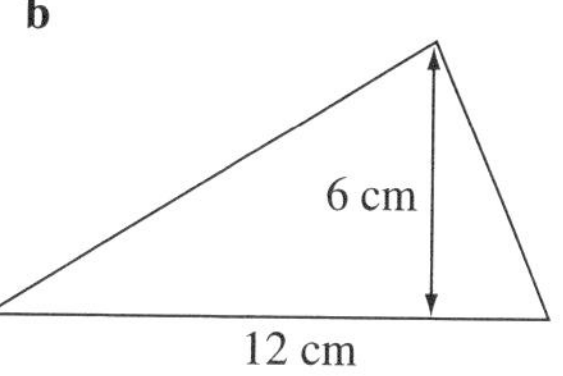

c

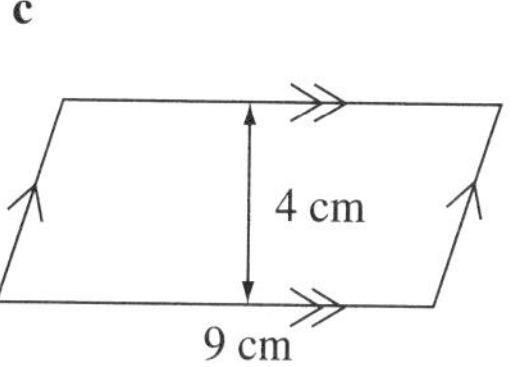

PS 4 A square has the same area as this parallelogram.

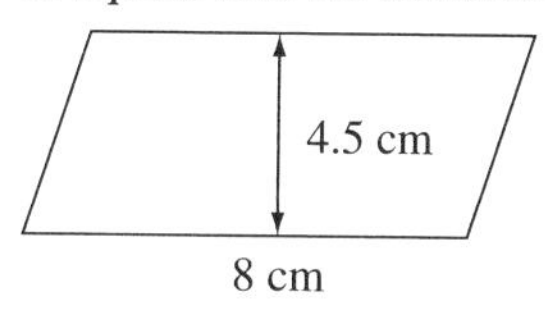

What is the perimeter of the square?

4.7 Area of a trapezium

HOMEWORK 4H

D

1 Calculate the perimeter and the area of each of these trapeziums.

a

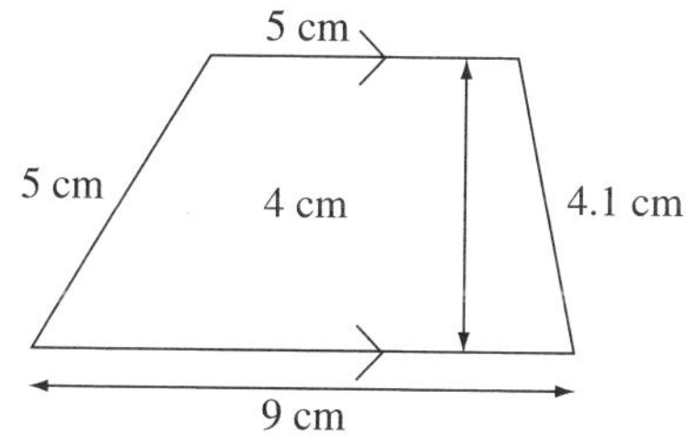

> **HINTS AND TIPS**
>
> Be careful not to use the slanting side as the height.

b

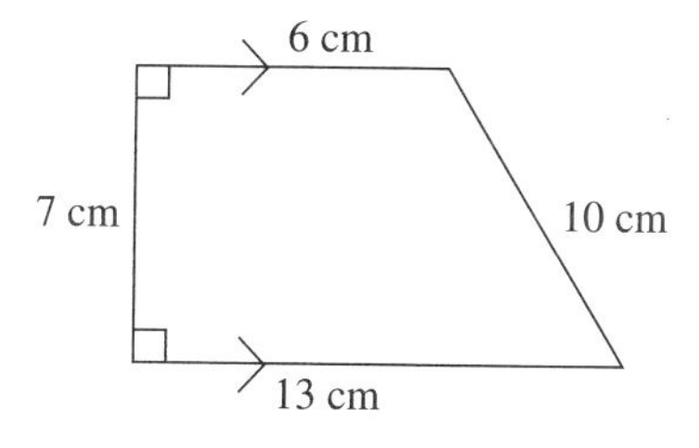

2 Calculate the area of each of these shapes.

a

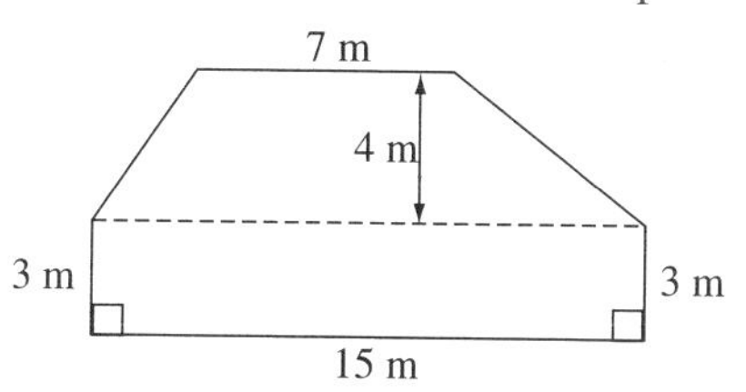

b

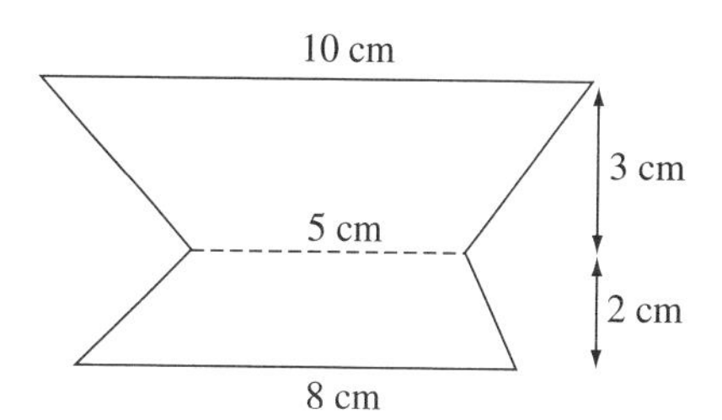

D

3 Calculate the area of the shaded part in each of these diagrams.

a

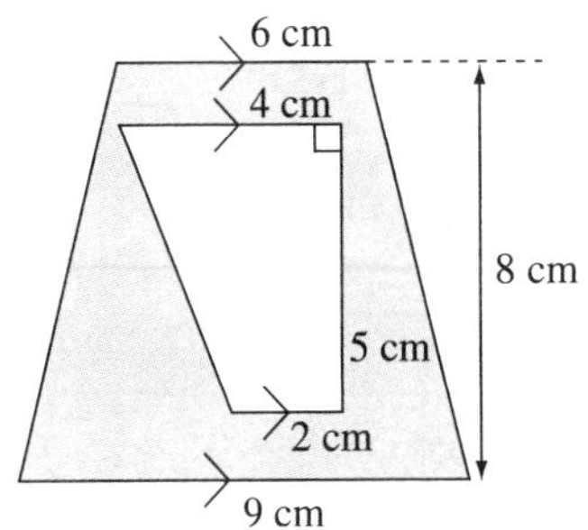

b

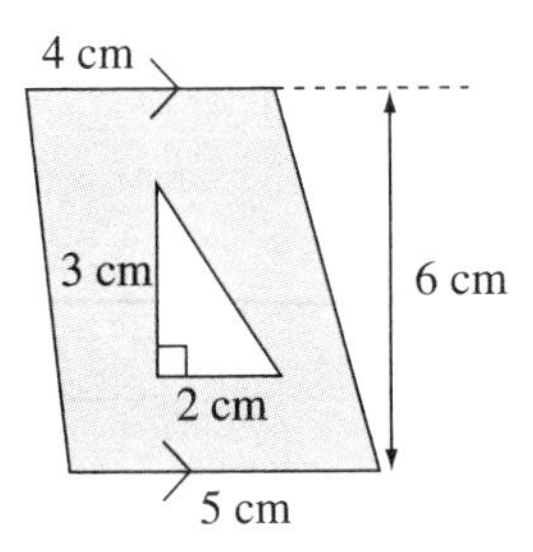

4 Which of the following shapes has the larger area?

a

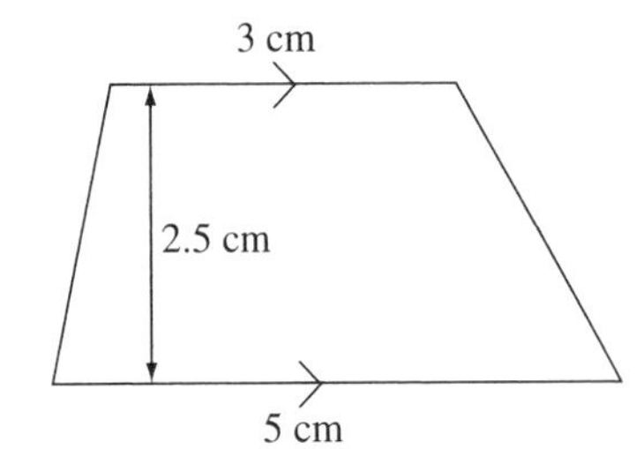

b

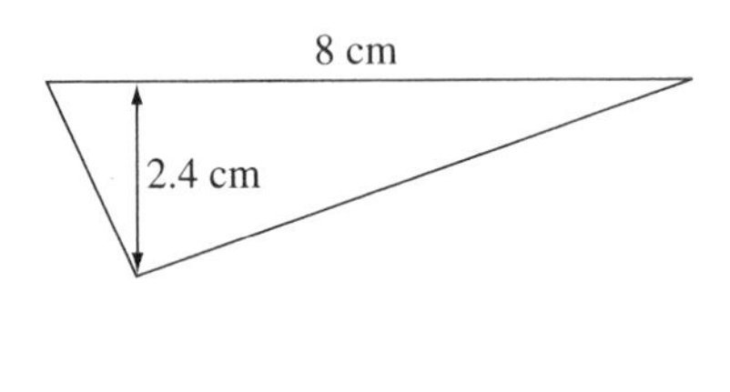

5 Priya is writing down her solution to this question.

Work out the area of this trapezium.

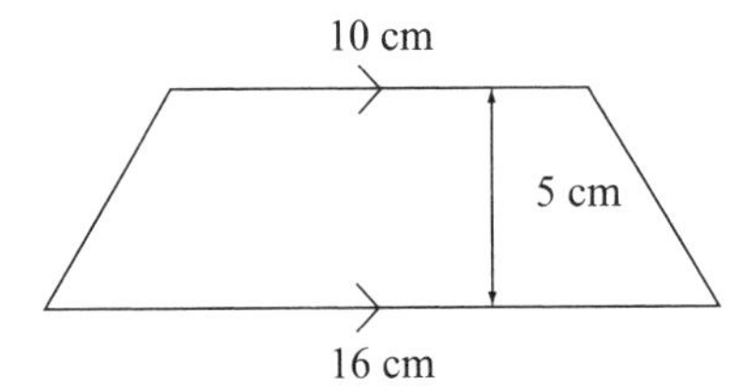

This is her answer.

Area $= \frac{1}{2}(10 + 16)$
$\times 5$

$= (5 + 16) \times 5$
$= 21 \times 5$

She has made two mistakes. Write out a correct solution to the question.

6 The side of a swimming pool is a trapezium, as shown in the diagram. Calculate its area.

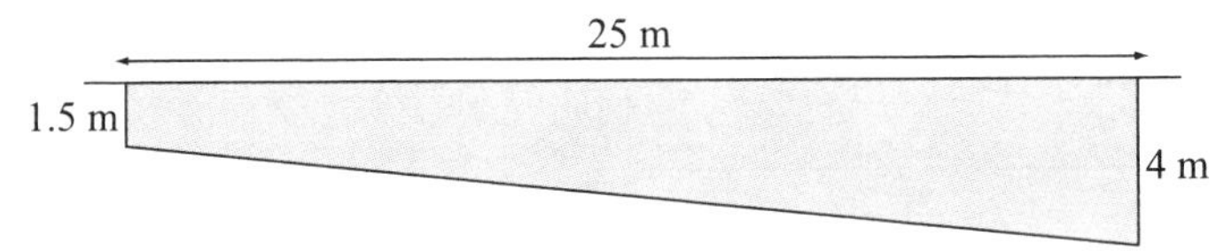

D

The area of this trapezium is 40 cm^2.
Work out possible values for a and b.

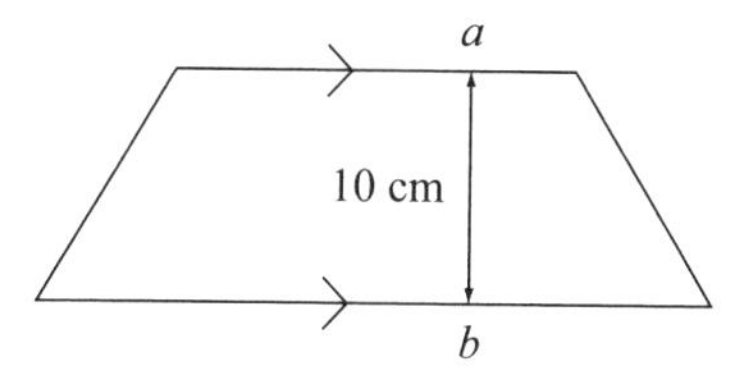

Problem-solving Activity

Pick's Theorem

Map makers and surveyors often need to calculate complex areas of land.

Here is a way to find the area of shapes drawn on a square dotty grid.

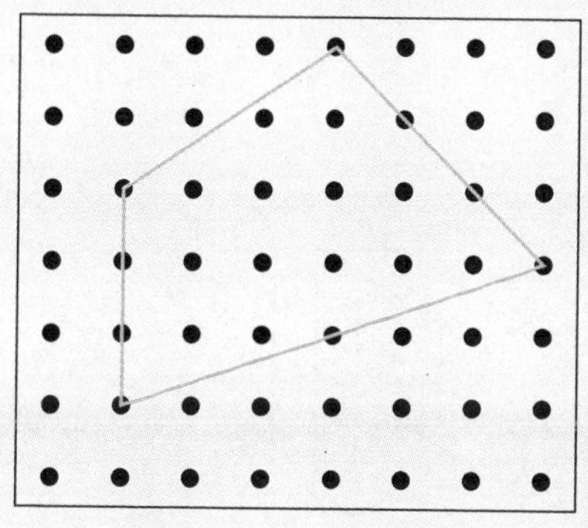

This quadrilateral has an area of $16\frac{1}{2}$ square units.

The perimeter of the quadrilateral passes through nine dots. Thirteen dots are contained within the perimeter of the quadrilateral.

Draw some quadrilaterals of different shapes and sizes on dotty paper. Make sure the **vertices** are all on dots on the paper. Investigate the connection between the area and the total number of dots inside and the total number of dots on the perimeter of the shape.

Then, from your findings, write down Pick's Theorem.

5 Geometry: Surface area and volume of 3D shapes

5.1 Units of volume

HOMEWORK 5A

Find the volume of each 3D shape if the edge of each cube is 1 cm.

1

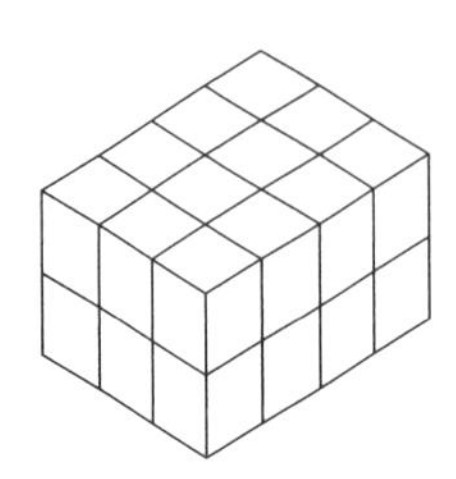

2

3

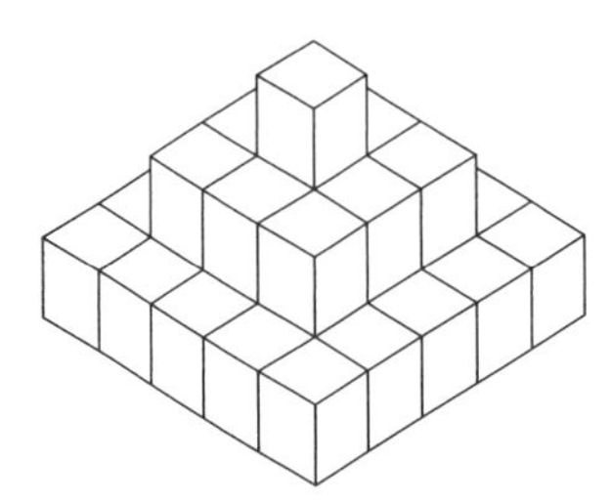

4

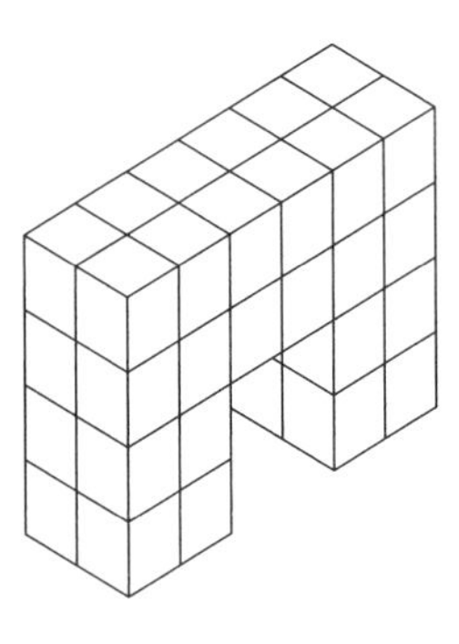

5.2 Surface area and volume of a cuboid

HOMEWORK 5B

Example Calculate the volume and surface area of this cuboid.

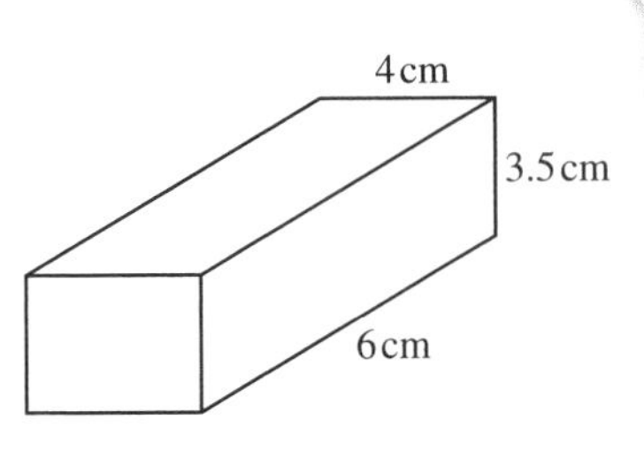

Volume = $6 \times 4 \times 3.5 = 84\ \text{cm}^3$

Surface area = $(2 \times 6 \times 4) + (2 \times 3.5 \times 4) + (2 \times 3.5 \times 6)$
$= 48 + 28 + 42 = 118\ \text{cm}^2$

1 Find **i** the volume and **ii** the surface area of each of these cuboids.

a

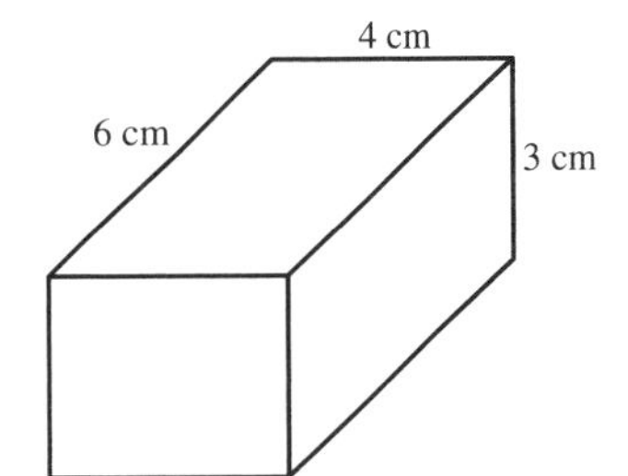

b

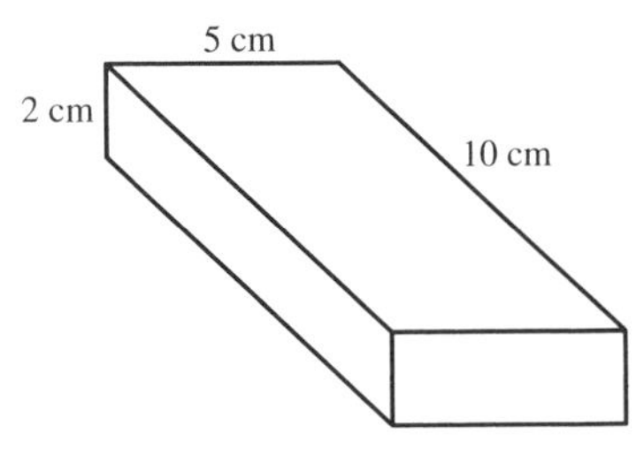

c

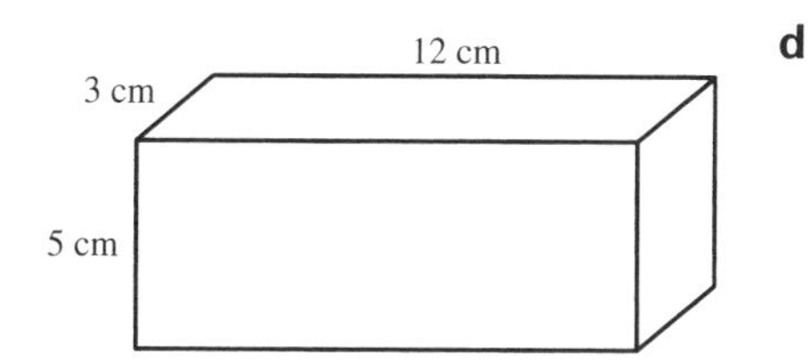

d

5 cm
5 cm
5 cm

2 Copy and complete the table, which shows the dimensions and volumes of four cuboids.

Length	Width	Height	Volume
4 cm	3 cm	2 cm	
	3 cm	3 cm	45 cm^3
8 cm		4 cm	160 cm^3
6 cm	6 cm		216 cm^3

3 Find the capacity (volume of a liquid or a gas) of a swimming pool that has the following dimensions: length 12 m, width 5 m and depth 1.5 m.

4 Find the volume of the cuboid in each of the following cases.

a The area of the base is 20 cm^2 and the height is 3 cm.

b The base has one side 4 cm, the other side 1 cm longer, and the height is 8 cm.

c The area of the top is 40 cm^2 and the depth is 3 cm.

FM 5 Safety notices in a hostel state that, where young people are sleeping, there should be at least 18 m^3 for each person in the room.
A dormitory in this hostel is 15 metres long, 12 metres wide and 3.5 metres high.
What is the largest number of young people who can safely sleep in this dormitory?

PS 6 What is the smallest surface area of a cuboid that has a volume of 512 cm^3?

AU 7 A cuboid has volume 216 cm^3 and a total surface area of 21 cm^2.
Is it possible for this cuboid to be a cube? Give a reason for your answer.

5.3 Surface area and volume of a prism

HOMEWORK 5C

1 For the prism below, calculate:

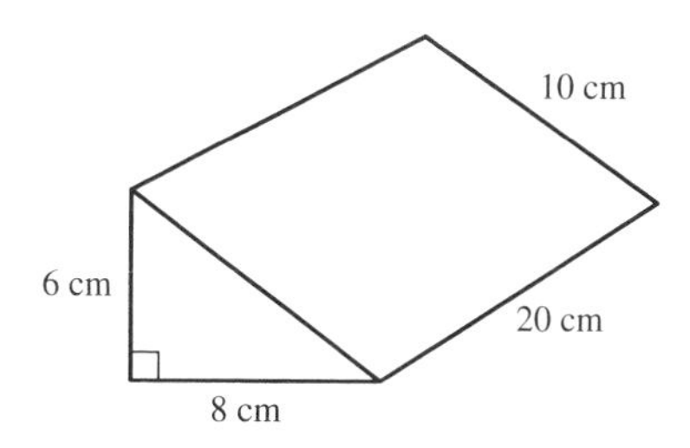

a its volume

b its total surface area.

C

2 For each prism shown, calculate **i** the area of the cross-section and **ii** the volume.

a

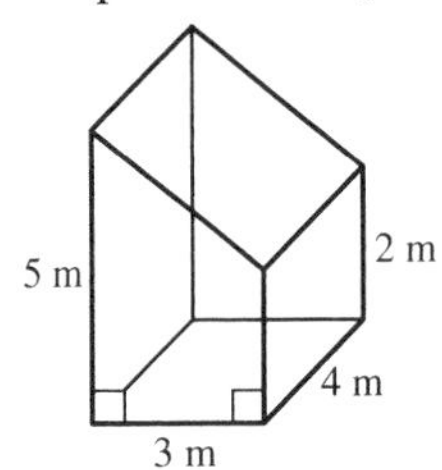

b

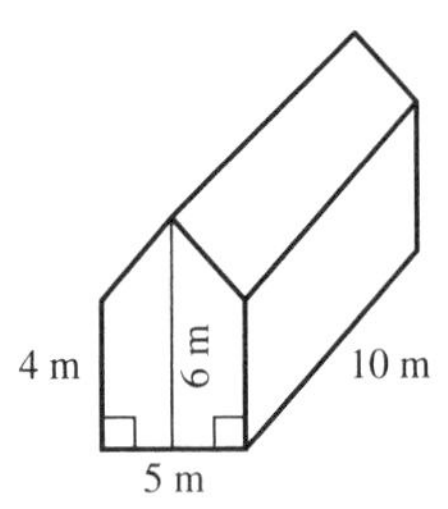

3 Calculate the weight of each prism.

a

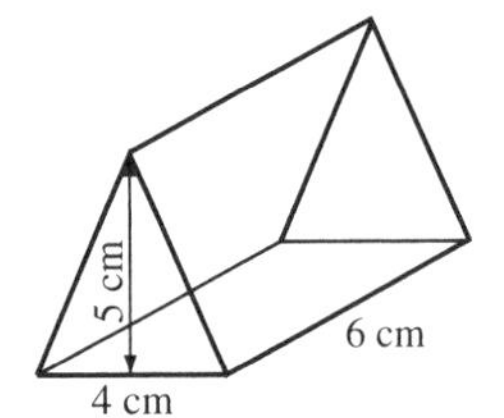

Density: 1 cm^3 weighs 3.13 g

b

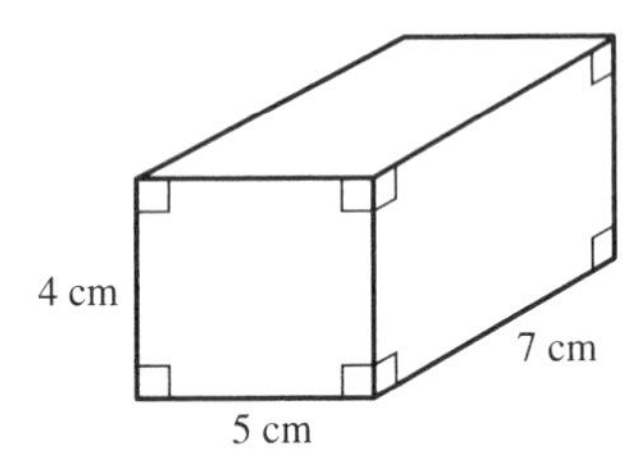

Density: 1 cm^3 weighs 1.35 g

FM **4** A 100 m trench is to be made for a construction job. It is in the shape of a trapezium that is 2.4 m wide at the top, 1.9 m wide at the bottom and 1.6 m deep.

a How much earth is to be removed?

b One lorry can carry a maximum load of 15 tonnes. 1 cm^3 of earth weighs 2.5 g. How many lorry loads will be needed to transport the earth away?

PS **5** A girl builds 27 cubes, each of edge 2 cm, into a single large cube.
How many more 2-cm cubes would she need to build a larger cube with edge 2 cm longer than the first one?

AU **6** Imagine you had a large glass bottle and you wished to mark on the outside the levels 1 litre, 2 litres, 3 litres, 4 litres, etc.
Explain how you would do this if you only had two measuring jugs – one 2-litre jug and one 5-litre jug.

5.4 Volume of a cylinder

HOMEWORK 5D

Example Calculate the volume of a cylinder with a radius of 4 cm and a height of 10 cm.

Volume = $\pi r^2 h = \pi \times 4^2 \times 10 = 502.7 \text{ cm}^3$ (to one decimal place).

C

1 Calculate the volume of each of these cylinders. Give your answers to one decimal place.

a Base radius 5 cm and a height of 7 cm.

b Base radius 10 cm and a height of 8 cm.

c Base diameter of 12 cm and a height of 20 cm.

d Base diameter of 9 cm and a height of 9 cm.

C

2 Find the volume of each of these cylinders. Give your answers to one decimal place.

a

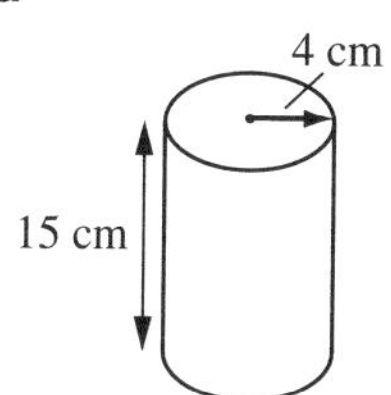

b

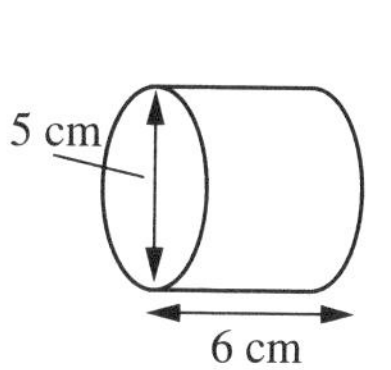

c

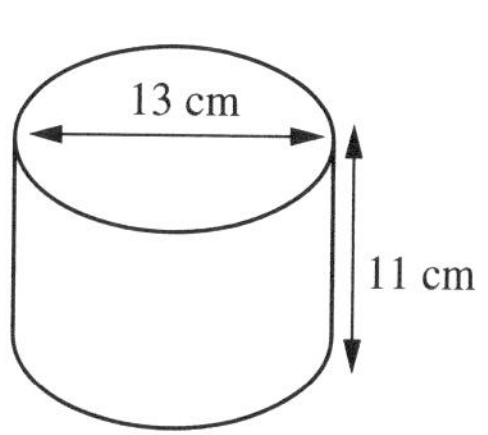

3 What is the weight of a solid iron bar 40 cm long with a radius of 2 cm? 1 cm^3 of iron weighs 8 g. Give your answer in kilograms.

4 Give the answers to this question in terms of π.

a What is the volume of a cylinder with a radius of 4 cm and a height of 11 cm?

b What is the volume of a cylinder with a diameter of 16 cm and a height of 18 cm?

FM **5** Andrea's mum is hosting a dinner party and is investigating how much wine to order.
A 'Tissan' wine glass is cylindrical in shape, with a radius of 1.5 cm and a height of 5 cm.
'Red Vin' wine can be bought in a box that is a cuboid shape, measuring 30 cm by 15 cm by 18 cm.
Calculate:

a the amount of Red Vin wine in a full box.

b the capacity of a Tissan wine glass.

c how many Tissan wine glasses can be 'half filled' from the box of Red Vin wine.

d There are going to be 30 guests at the party. If Andrea's mum serves each guest three half-filled glasses of wine, how many boxes of Red Vin should she order?

FM PS **6** A tunnel is cut through a hillside in the shape of a semicircle. The diameter of the semicircle is 15 m and the length of the tunnel is 250 m.
One lorry can take away 8 m^3 of waste.
How many lorry loads are needed to move all the waste that is produced from cutting through the hillside?

AU **7** A roll of paper is delivered to a printer. It is 80 cm in diameter and the paper is wound onto a wooden cylindrical block 10 cm in diameter. The paper is 0.004 cm thick.
What length of paper is there on the roll?

Functional Maths Activity

Baking cakes

Dahlia has a recipe for a sponge cake.
It uses 100 grams of flour.
In a circular tin with a diameter of 18 cm, it makes a sponge 3 cm high.

1 She wants to make a sponge cake of the same height in a circular tin with a diameter of 25 cm.
She has been told that this needs 200 grams of flour, which is twice as much.
Do you think this is correct? Give a reason for your answer.

2 She also has a square tin with a side of 16 cm. She wants to know how much flour she needs to make a sponge 3 cm high.
Help her to find out how much flour she will need.

3 What size square tin would need 200 grams of flour to make a sponge 3 cm high?

Algebra: More complex equations and formulae

6.1 Solving equations with brackets

HOMEWORK 6A

Example Solve $3(2x - 7) = 15$.

First multiply out the bracket to get $6x - 21 = 15$
Add 21 to both sides $6x = 36$
Divide both sides by 6 $x = 6$

D

1 Solve each of the following equations. Some of the answers may be decimals or negative numbers. Remember to check that each answer works in the original equation. Use your calculator if necessary.

a $2(x + 1) = 8$ **b** $3(x - 3) = 12$ **c** $3(t + 2) = 9$ **d** $2(x + 5) = 20$
e $2(2y - 5) = 14$ **f** $2(3x + 4) = 26$ **g** $4(3t - 1) = 20$ **h** $2(t + 5) = 6$
i $2(x + 4) = 2$ **j** $2(3y - 2) = 5$ **k** $4(3k - 1) = 11$ **l** $5(2x + 3) = 26$

AU 2 Mike has been asked to solve the equation $a(bx + c) = 60$.
Mike knows that the values of a, b and c are 2, 4 and 5, but he doesn't know which is which.
He also knows that the answer is an even number.
What are the correct values of a, b and c?

PS 3 As the class are coming in for the start of a maths lesson, the teacher is writing some equations on the board.
So far she has written:
$5(2x + 3) = 13$
$2(5x + 3) = 13$
Zak says 'That's easy. Both equations have the same solution, $x = 2$.'
Is Zak correct? If not, what mistake has he made? What are the correct answers?

6.2 Equations with the variable on both sides

HOMEWORK 6B

Example Solve $5x + 4 = 3x + 10$.

Subtract $3x$ from both sides $2x + 4 = 10$
Subtract 4 from both sides $2x = 6$
Divide both sides by 2 $x = 3$

D

1 Solve each of the following equations.

a $2x + 1 = x + 3$ **b** $3y + 2 = 2y + 6$ **c** $5a - 3 = 4a + 4$
d $5t + 3 = 3t + 9$ **e** $7p - 5 = 5p + 3$ **f** $6k + 5 = 3k + 20$
g $6m + 1 = m + 11$ **h** $5s - 1 = 2s - 7$ **i** $4w + 8 = 2w + 8$
j $5x + 5 = 3x + 10$

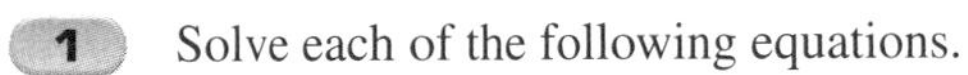

PS 2

Ernie: I am thinking of a number, I multiply it by 3 and subtract 6.

Eric: I am thinking of a number, I multiply it by 5 and add 2.

Eric and Ernie find that they both thought of the same number and both got the same final answer.

What number did they think of?

3 Solve each of the following equations.

a $5(t - 2) = 4t - 1$ **b** $4(x + 2) = 2(x + 1)$ **c** $5p - 2 = 5 - 2p$

d $2(2x + 3) = 3(x - 4)$

AU 4 The triangle shown is isosceles.

What is the perimeter of the triangle?

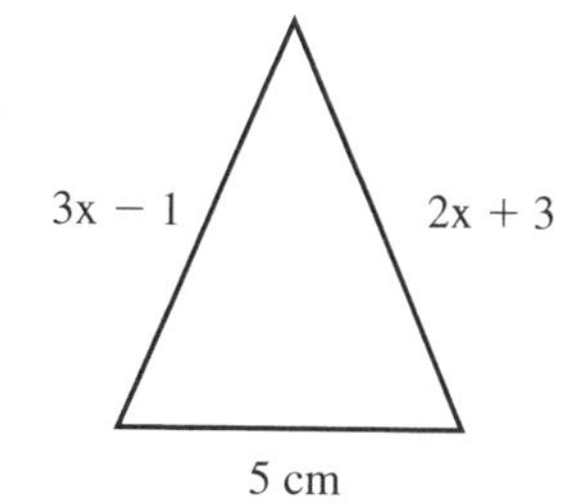

6.3 Rearranging formulae

HOMEWORK 6C

1 $y = 2x + 3$ Make x the subject.

2 $v = u - 10$ Make u the subject.

3 $T = 2 + 3y$ Make y the subject.

4 $p = q^2$ Make q the subject.

5 $p = \frac{q}{L}$ Make q the subject.

6 $2a = 5b + 1$ Make b the subject.

7 A rocket is fired vertically upwards with an initial velocity of u metres per second. After t seconds the rocket's velocity, v metres per second, is given by the formula $v = u + 10t$.

a Calculate v when $u = 120$ and $t = 6$

b Rearrange the formula to make t the subject

c Calculate t when $u = 20$ and $v = 100$

C

FM 8 A restaurant has a large oven that can cook up to 10 chickens at a time.
The restaurant uses the following formula for the length of time it tales to cook n chickens:

$T = 10n + 55$

A large party is booked for a chicken dinner at 7 pm. They will need eight chickens between them.

a It takes 15 minutes to get the chickens out of the oven and prepare them for serving. At what time should the eight chickens go into the oven?

b The next day, another large party is booked in.

i Rearrange the formula to make n the subject.

ii The party is booked in at 8 pm and the chef calculates she will need to put the chickens in the oven at 5.50 pm. How many chickens does the party need?

AU 9 Kern notices that the price of six coffees is 90 pence cheaper than the price of nine teas.
Let the price of a coffee be x pence and the price of a tea be y pence.

a Express the cost of a tea, y, in terms of the price of a coffee, x.

b If the price of a coffee is £1.20, how much is a tea?

PS 10 Distance, speed and time are connected by the formula Distance = Speed × Time.
A delivery driver drove 90 miles at an average speed of 60 miles per hour.
On the return journey, he was held up at some road works for 30 minutes.
What was his average speed on the return journey?

Problem-solving Activity

Number problems

What number is being described here?

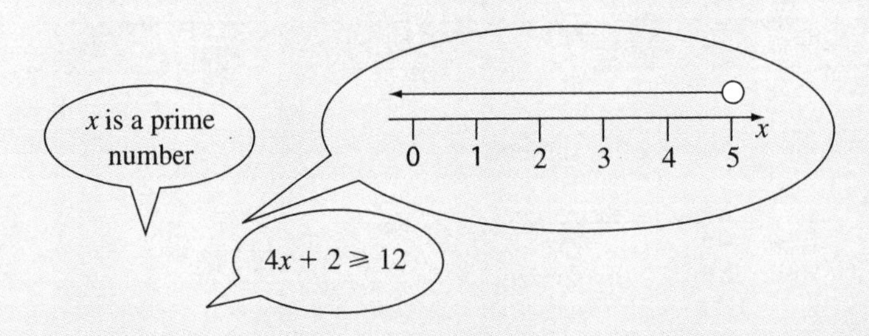

Now choose a number and write three of your own statements to describe it.

7 Number: Further arithmetic with fractions

7.1 Place value and ordering numbers

HOMEWORK 7A

1 Write the value of each underlined digit.

a $5\underline{7}6$ **b** $37\underline{4}$ **c** $\underline{6}89$ **d** $\underline{4}785$ **e** $300\underline{7}$

f $7\underline{6}08$ **g** $354\underline{2}$ **h** $1\underline{2}\ 745$ **i** $\underline{8}7\ 409$ **j** $\underline{7}\ 777\ 777$

2 Write each of the following using just words.

a 7245 **b** 9072 **c** 29 450 **d** 2 760 000 **e** 5 800 000

3 Write each of the following using digits only.

a Eight thousand and five hundred **b** Forty two thousand and forty two

c Six million **d** Five million and five

4 Write these numbers in order, putting the **smallest** first.

a 31, 20, 14, 22, 8, 25, 30, 12

b 159, 155, 176, 167, 170, 168, 151, 172

c 2100, 2070, 2002, 1990, 2010, 1998, 2000, 2092

5 Write these numbers in order, putting the **largest** first.

a 49, 62, 75, 57, 50, 72

b 988, 1052, 999, 1010, 980, 1007

c 4567, 4765, 4675, 4576, 4657, 4756

FM 6 Here are the distances from home to five seaside towns.

Skegness (86 miles)
Rhyl (115 miles)
Great Yarmouth (166 miles)
Scarborough (80 miles)
Blackpool (105 miles)

a Which place is the furthest from home?

b Which place is the nearest to home?

PS 7 Using each of the digits 7, 8 and 9 only once in each number:

a write as many three-digit numbers as you can.

b which of your numbers is the smallest?

c which of your numbers is the largest?

PS 8 Write down in order of size, largest first, all the two-digit numbers that can be made using 2, 4 and 6. (Each digit can be repeated.)

9 Copy each of these sentences, writing the numbers in words.

a The diameter of the Earth at the equator is 12 756 kilometres.

b The Moon is approximately 238 000 miles from the Earth.

c The greatest distance of the Earth from the Sun is 94 600 000 miles.

AU 10 Using each of the digits 1, 5, 6 and 9, make a four-digit even number greater than eight thousand.

G

F

7.2 Adding and subtracting simple fractions

HOMEWORK 7B

Example

a $\frac{5}{12} + \frac{4}{12} = \frac{9}{12}$

b $\frac{7}{10} - \frac{3}{10} = \frac{4}{10}$

G

1 Calculate each of the following.

a $\frac{1}{4}+\frac{1}{4}$ b $\frac{2}{5}+\frac{1}{5}$ c $\frac{3}{7}+\frac{2}{7}$ d $\frac{5}{8}+\frac{1}{8}$ e $\frac{5}{6}+\frac{2}{6}$

f $\frac{4}{9}+\frac{4}{9}$ g $\frac{3}{10}+\frac{4}{10}$ h $\frac{2}{5}+\frac{2}{5}$ i $\frac{4}{12}+\frac{1}{12}$ j $\frac{5}{20}+\frac{7}{20}$

2 Calculate each of the following.

a $\frac{4}{5}-\frac{2}{5}$ b $\frac{5}{8}-\frac{1}{8}$ c $\frac{6}{7}-\frac{2}{7}$ d $\frac{8}{10}-\frac{3}{10}$ e $\frac{5}{6}-\frac{3}{6}$

f $\frac{7}{9}-\frac{3}{9}$ g $\frac{7}{8}-\frac{1}{8}$ h $\frac{4}{9}-\frac{2}{9}$ i $\frac{7}{12}-\frac{5}{12}$ j $\frac{11}{20}-\frac{3}{20}$

PS **3**
a Draw two diagrams to show $\frac{4}{8}$ and $\frac{2}{8}$.
b Show on your diagrams that $\frac{4}{8} = \frac{1}{2}$ and $\frac{2}{8} = \frac{1}{4}$.
c Use the above information to write down the answers to each of the following.
i $\frac{1}{2}+\frac{1}{8}$ ii $\frac{1}{2}+\frac{3}{8}$ iii $\frac{1}{4}+\frac{1}{8}$ iv $\frac{3}{4}+\frac{1}{8}$ v $\frac{1}{2}-\frac{1}{8}$ vi $\frac{1}{2}-\frac{3}{8}$ vii $\frac{1}{4}-\frac{1}{8}$ viii $\frac{3}{4}-\frac{1}{8}$

AU **4** Copy the diagrams below, shade them to show the working for this question and then write down the answer.

$\frac{1}{6}$ + $\frac{1}{3}$ = ...

[| | | | |] + [| | | | |] = [| | | | |]

F

FM **5** There are ten people at a bus stop.
Six get on a bus but four other people arrive.
a How many people have been at the bus stop?
b What fraction of all these people are now at the bus stop?
c A man is giving away 100 free newspapers at the bus stop.
Every 10 minutes 15 people arrive at the bus stop and half of them take a paper.
How long will it take him to give away all the papers?

7.3 Improper fractions and mixed numbers

HOMEWORK 7C

F

1 Change each of these top-heavy fractions into a mixed number.

a $\frac{5}{2}$ b $\frac{5}{3}$ c $\frac{7}{4}$ d $\frac{11}{3}$ e $\frac{9}{2}$ f $\frac{13}{4}$

g $\frac{11}{5}$ h $\frac{10}{4}$ i $\frac{14}{6}$ j $\frac{17}{8}$ k $\frac{17}{10}$ l $\frac{26}{8}$

m $\frac{12}{4}$ n $\frac{20}{5}$ o $\frac{60}{10}$

2 Change each of these mixed numbers into a top-heavy fraction.

a $1\frac{1}{2}$ b $2\frac{1}{4}$ c $2\frac{1}{3}$ d $4\frac{1}{2}$ e $3\frac{2}{3}$ f $1\frac{3}{4}$

g $2\frac{1}{5}$ h $2\frac{3}{8}$ i $3\frac{2}{5}$ j $4\frac{3}{5}$ k $5\frac{3}{8}$ l $4\frac{3}{7}$

m $5\frac{4}{9}$ n $4\frac{5}{12}$ o $7\frac{7}{10}$

3 Check your answers to Questions **1** and **2** using the fraction buttons on your calculator.

AU **4** Which of these improper fractions is the biggest?

$\frac{7}{2}$ $\frac{10}{3}$ $\frac{17}{5}$

Show working to justify your answer.

PS **5** Find a mixed number that is greater than $\frac{59}{8}$ and less than $\frac{53}{7}$.

PS **6** Here is a list of numbers.

2 5 9 19

Using one of these numbers for the numerator and one for the denominator, find an improper fraction with a value between 1 and 2.

7.4 Adding and subtracting fractions with the same denominator

HOMEWORK 7D

Example 1 $\frac{5}{9} + \frac{7}{9} = \frac{10}{9} = \frac{4}{3} = 1\frac{1}{3}$ (Cancel down and change to a mixed number.)

Example 2 $\frac{2}{3} + \frac{1}{5} = \frac{10}{15} + \frac{3}{15} = \frac{13}{15}$ (Use equivalent fractions to make the denominators the same.)

1 Copy and complete each of these additions.

a $\frac{2}{7} + \frac{4}{7}$ **b** $\frac{4}{9} + \frac{1}{9}$ **c** $\frac{2}{5} + \frac{2}{5}$ **d** $\frac{2}{11} + \frac{3}{11}$ **e** $\frac{3}{13} + \frac{4}{13}$

2 Copy and complete each of these subtractions.

a $\frac{3}{7} - \frac{2}{7}$ **b** $\frac{7}{9} - \frac{2}{9}$ **c** $\frac{7}{11} - \frac{2}{11}$ **d** $\frac{8}{13} - \frac{2}{13}$ **e** $\frac{4}{5} - \frac{1}{5}$

3 Calculate each of these additions. Remember to cancel down.

a $\frac{3}{8} + \frac{1}{8}$ **b** $\frac{3}{10} + \frac{5}{10}$ **c** $\frac{5}{12} + \frac{1}{12}$ **d** $\frac{1}{9} + \frac{5}{9}$ **e** $\frac{2}{15} + \frac{7}{15}$

4 Copy and complete each of these subtractions. Remember to cancel down.

a $\frac{5}{8} - \frac{4}{8}$ **b** $\frac{9}{10} - \frac{3}{10}$ **c** $\frac{5}{9} - \frac{2}{9}$ **d** $\frac{9}{10} - \frac{3}{10}$ **e** $\frac{7}{8} - \frac{1}{8}$

5 Calculate each of these additions. Remember to use equivalent fractions.

a $\frac{1}{3} + \frac{1}{2}$ **b** $\frac{2}{5} + \frac{3}{10}$ **c** $\frac{1}{4} + \frac{5}{12}$ **d** $\frac{3}{5} + \frac{1}{4}$ **e** $\frac{3}{4} + \frac{2}{3}$

6 Calculate each of these subtractions.

a $\frac{5}{8} - \frac{1}{8}$ **b** $\frac{7}{10} - \frac{3}{10}$ **c** $\frac{11}{12} - \frac{3}{4}$ **d** $\frac{2}{3} - \frac{1}{2}$ **e** $\frac{9}{10} - \frac{1}{5}$

7 At a cricket match, $\frac{9}{10}$ of the crowd were men. What fraction of the crowd were women?

8 An iceberg shows $\frac{1}{9}$ of its mass above sea level. What fraction of it is below sea level?

9 A petrol gauge shows that a tank is $\frac{7}{12}$ full. What fraction of the tank is empty?

FM **10** David spends $\frac{1}{4}$ of his pocket money on bus fares, $\frac{1}{3}$ on magazines and saves the rest. He wants to save half of his pocket money. Does he succeed?

11 In a local election Mr Weeks received $\frac{2}{5}$ of the total votes, Ms Meenan received $\frac{1}{4}$ and Mr White received the remainder. What fraction of the total votes did Mr White receive?

AU **12** On a certain day at a busy railway station, $\frac{7}{10}$ of the trains arriving were on time, $\frac{1}{6}$ were late by 10 minutes or less and the rest were late by more than 10 minutes. What fraction of the trains arrived late by more than 10 minutes?

7.5 Multiplying and dividing fractions

HOMEWORK 7E

E

1 Work out each of these multiplications.

a $\frac{1}{2} \times \frac{1}{2}$ **b** $\frac{1}{3} \times \frac{1}{5}$ **c** $\frac{1}{4} \times \frac{1}{3}$

d $\frac{3}{4} \times \frac{1}{2}$ **e** $\frac{1}{3} \times \frac{3}{5}$ **f** $\frac{2}{3} \times \frac{1}{2}$

g $\frac{4}{5} \times \frac{1}{2}$ **h** $\frac{5}{6} \times \frac{1}{5}$ **i** $\frac{3}{8} \times \frac{2}{3}$

j $\frac{3}{10} \times \frac{5}{6}$

FM PS **2** A printer is cutting sheets of paper down to smaller sizes.
He cuts each sheet into quarters and then cuts each quarter in half.

a What fraction of the original sheet is each new sheet?

b Each small sheet is sold for 10p. How many large sheets should be cut up to make £20?

D

AU **3** There are 270 000 people living in Rotherham. One-sixth of these are aged over 60. Of the over-60s, two-thirds are women.

a What fraction of the whole population are women over 60?

b How many woman over 60 live in Rotherham?

Functional Maths Activity

Organising an activity holiday

On an activity holiday there are **180** students and you want to organise them in to groups for different activities.
The activity leaders give you the following facts:

Watersports (windsurfing and sailing)
All students take one watersport, except for one group of 26 students who do a rope course instead.

Windsurfing
Half of all students do windsurfing.
They are all taught at the same time.
There are three equal-sized groups.

Sailing
There are three groups.
The smallest group has one-quarter of the students doing sailing.
The other two groups have the same number of students in each.

Abseiling
Two-thirds of all students do abseiling.
There are five equal-sized groups.

Rock climbing
Two-fifths of all students do rock climbing.
One group has 21 students and one group has 18 students.
There are two other groups. Two-thirds of the remaining students are in one group.

Copy and complete the table for the activity groups. Remember that there are 180 students in total.
Choose your own rules for mountain biking.
Decide the number of groups and work out the number in each group so that no mountain biking group exceeds 30 students.

Activity	**Number in each group**			
Windsurfing				
Sailing				
Abseiling				
Rock climbing				
Mountain biking				

Now copy and complete this table to give the fraction of the total students in each group.

Activity	**Fraction of total students in each group**			
Windsurfing				
Sailing				
Abseiling				
Rock climbing				
Mountain biking				

8 Number: Properties of number

8.1 Multiples of whole numbers

HOMEWORK 8A

G

1 Write out the first five multiples of:

a 4 **b** 6 **c** 8 **d** 12 **e** 15.

Remember: the first multiple is the number itself.

2 From the list of numbers below

28 19 36 43 64 53 77 66 56 60 15 29 61 45 51

write down those that are:

a multiples of 4 **b** multiples of 5 **c** multiples of 8 **d** multiples of 11.

3 Use your calculator to see which of the numbers below are:

a multiples of 7 **b** multiples of 9 **c** multiples of 12.

225 252 361 297 162 363 161 289 224 205 312 378 315 182 369

PS **4** Find the biggest number smaller than 200 that is:

a a multiple of 2 **b** a multiple of 4 **c** a multiple of 5 **d** a multiple of 8

e a multiple of 9.

PS **5** Find the smallest number that is a multiple of 3 and bigger than:

a 10 **b** 100 **c** 1000 **d** 10 000 **e** 1 000 000 000.

6 There are 12 sweets in a bag. There are 96 sweets ready to put into bags.
Will all the bags be full?
Give a reason for your answer.

7 48 people are at a wedding reception. The tables are arranged so that the same number of people sit at each table.
How many people sit at each table?
Give **two** possible answers.

8 Here is a list of numbers.

4 9 10 12 14 20

a From the list, write down a multiple of 7.

b From the list, write down a multiple of 6.

c From the list, write down a multiple of both 4 and 5.

F

PS **9** Find the lowest odd number that is a multiple of 9 and a multiple of 15.

8.2 Factors of whole numbers

HOMEWORK 8B

Example Find the factors of 32.

Look for the pairs of numbers which make 32 when multiplied together. These are

$1 \times 32 = 32$, $2 \times 16 = 32$ and $4 \times 8 = 32$. So the factors of 32 are 1, 2, 4, 8, 16, 32.

G

1 What are the factors of each of these numbers?

a 12 **b** 13 **c** 15 **d** 20 **e** 22
f 36 **g** 42 **h** 48 **i** 49 **j** 50

2 Use your calculator to find the factors of each of these numbers.

a 100 **b** 111 **c** 125 **d** 132 **e** 140

PS **3** All the numbers in **a** to **j** are divisible by 11. Use your calculator to divide each one by 11 and then write down the answer. What do you notice?

a 143 **b** 253 **c** 275 **d** 363 **e** 462
f 484 **g** 561 **h** 583 **i** 792 **j** 891

FM **4** Fred wants to pack 18 items into boxes so that there are exactly the same number of items in each box. How many ways can he do this?

5 Here is a list of numbers.

3 6 8 10 13

a From the list, write down a factor of 32.
b From the list, write down a factor of 20.
c From the list, write down a factor of both 26 and 39.

AU **6** Here are five numbers

15 20 24 27 30

Use factors to explain why 20 could be the odd one out.

PS **7** Find the highest even number that is a factor of 30 and a factor of 42.

F

8.3 Prime numbers

HOMEWORK 8C

1 Write down all the prime numbers less than 40.

2 Which of these numbers are prime?

43 47 49 51 54 57 59 61 65 67

3 This is a number pattern to generate odd numbers.

Line 1 $2 - 1 = 1$
Line 2 $2 \times 2 - 1 = 3$
Line 3 $2 \times 2 \times 2 - 1 = 7$

a Work out the next three lines of the pattern.
b Which lines have answers that are prime numbers?

PS **4** Using the rules for recognising multiples, decide which of these numbers are not prime.

39 41 51 71 123

AU **5** When two different prime numbers are multiplied together the answer is 91.
What are the two prime numbers?

6 **a** Write down two prime numbers with a difference of 6.
b Write down two more prime numbers with a difference of 6.

FM **7** A mechanic has a set of 23 spanners.
Is it possible to put them in a toolbox so that he has the same number of spanners in each part of his box?
Explain your answer.

E

8.4 Square numbers

HOMEWORK 8D

1 Write down the first ten square numbers.

PS 2 Here is a number pattern.

$2 \times 0 + 1 = 1$
$3 \times 1 + 1 = 4$
$4 \times 2 + 1 = 9$

a Write down the next three lines in the pattern.
b Describe what you notice about the answers to each line of the pattern.

3 Write down the answer to each of the following. You will need to use your calculator.

a 5^2 **b** 15^2 **c** 25^2 **d** 35^2 **e** 45^2
f 55^2 **g** 65^2 **h** 75^2 **i** 85^2 **j** 95^2

Describe any pattern you notice.

4 **a** Write down the value of 11^2.
b Estimate the value of 10.5^2.

5 How much do 15 rulers at 15 pence each cost?

6 A builder buys 60 bricks for 60 pence each.
She has £40. How many extra bricks can she afford to buy?

7 In a warehouse, books are stored on shelves in piles of 20.
How many books are on two shelves, if there are ten piles of books on each shelf?

HOMEWORK 8E

1 Write down the first five multiples of:

a 5 **b** 7 **c** 16 **d** 25 **e** 30.

Remember: the first multiple is the number itself.

2 Write down all the factors of each of these numbers.

a 18 **b** 25 **c** 28 **d** 35 **e** 40

3 Write down the first three numbers that are multiples of both:

a 2 and 5 **b** 3 and 4 **c** 5 and 6 **d** 4 and 6 **e** 8 and 10.

4 In a prize draw, raffle tickets are numbered from 1 to 100.
A prize is given if a ticket drawn is a multiple of 10 or a multiple of 15.
Which ticket holders will receive two prizes?

AU 5 Here is a number pattern using square numbers.

$1^2 - 0^2 = 1$
$2^2 - 1^2 = 3$
$3^2 - 2^2 = 5$
$4^2 - 3^2 = 7$

a Write down the next three lines in the pattern.
b What do you think is the answer to $21^2 - 20^2$?
Explain your answer.

6 From the list of numbers below:

4 6 7 10 13 16 21 23 25 28 34 37 40 49 50

write down those that are:

a prime numbers;

b square numbers.

AU 7 Here are four numbers.

3 12 25 36

Copy and complete the table by putting the numbers in the correct boxes.

	Square number	**Factor of 24**
Odd number		
Multiple of 6		

AU 8 Use these four number cards to make a square number.

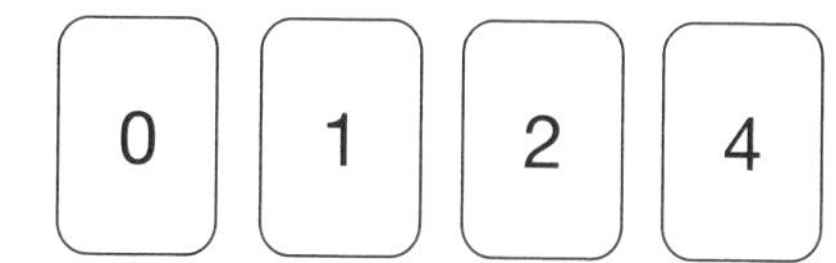

8.5 Square roots

HOMEWORK 8F

1 Write down the positive square root of each of these numbers.

a 64 **b** 25 **c** 49 **d** 81 **e** 16
f 36 **g** 100 **h** 121 **i** 144 **j** 400

2 Write down the answer to each of the following. You will need to use your calculator.

a $\sqrt{225}$ **b** $\sqrt{289}$ **c** $\sqrt{441}$ **d** $\sqrt{625}$ **e** $\sqrt{1089}$
f $\sqrt{1369}$ **g** $\sqrt{3136}$ **h** $\sqrt{6084}$ **i** $\sqrt{40\,804}$ **j** $\sqrt{110\,889}$

PS 3 Here is a number pattern using square roots and square numbers.

$$\sqrt{1} = 1$$
$$\sqrt{1} + \sqrt{4} = 3$$
$$\sqrt{1} + \sqrt{4} + \sqrt{9} = 6$$

a Write down the next three lines in the pattern.

b Describe any pattern you notice in the answers.

AU 4 Put these in order starting with the smallest value.

2^2 $\sqrt{20}$ $\sqrt{10}$ 3^2

AU 5 Between which two consecutive whole numbers does the square root of 40 lie?

PS 6 A child has 125 square tiles which she is arranging into square patterns.
How many tiles will be in the biggest square she can make?

FM 7 Square floor tiles are being fitted on a square kitchen floor.
Altogether it needs 121 tiles.
How many tiles are needed for each row?

8.6 Powers

HOMEWORK 8G

Example Work out 3^5.

$3^5 = 3 \times 3 \times 3 \times 3 \times 3 = 243$

E

1 Use your calculator to work out the value of each of the following.

a 2^3 **b** 4^3 **c** 7^3 **d** 10^3 **e** 12^3
f 3^4 **g** 10^4 **h** 2^5 **i** 10^6 **j** 2^8

PS 2 Use your calculator to work out the answers to the following powers of 11.

a 11^2 **b** 11^3 **c** 11^4

Describe any patterns you notice in your answers.
Does your pattern work for other powers of 11? Give a reason for your answer.

3

1	2	3	4	5	6	7	8	9
10	11	12	13	14	15	16	17	18
19	20	21	22	23	24	25	26	27
28	29	30	31	32	33	34	35	36

From the numbers above, write down:

a all the multiples of 7
b all the factors of 30
c all the prime numbers
d the square of 6
e the square root of 25
f the cube of 3.

4 A box is in the shape of a cube.
The height of the box is 30 centimetres.
To work out the volume of a cube, use the formula
Volume = (length of edge)^3
Work out the volume of the box.

D

AU 5 Write each number as a power of a different number.
The first one has been done for you.

a $16 = 2^4$
b $64 =$
c $27 =$
d $36 =$

8.7 Multiplying and dividing by powers of 10

HOMEWORK 8H

D

1 Evaluate the following.

a 3.5×100 **b** 2.15×10 **c** 6.74×1000 **d** 4.63×10
e 30.145×10 **f** 78.56×1000 **g** 6.42×10^2 **h** 0.067×10
i 0.085×10^3 **j** 0.798×10^5 **k** 0.658×1000 **l** 215.3×10^2
m 0.889×10^6 **n** 352.147×10^2 **o** 37.2841×10^3 **p** 34.28×10^6

2 Evaluate the following.

a $4538 \div 100$ **b** $435 \div 10$ **c** $76459 \div 1000$ **d** $643.7 \div 10$
e $4228.7 \div 100$ **f** $278.4 \div 1000$ **g** $246.5 \div 10^2$ **h** $76.3 \div 10$
i $76 \div 10^3$ **j** $897 \div 10^5$ **k** $86.5 \div 1000$ **l** $1.5 \div 10^2$
m $0.8799 \div 10^6$ **n** $23.4 \div 10^2$ **o** $7654 \div 10^3$ **p** $73.2 \div 10^6$

D

3 Evaluate the following.

a	400×300	**b**	50×4000	**c**	70×200	**d**	30×700
e	$(30)^2$	**f**	$(50)^3$	**g**	$(200)^2$	**h**	40×150
i	70×200	**j**	60×5000	**k**	30×250	**l**	700×200

4 Evaluate the following.

a	$4000 \div 800$	**b**	$9000 \div 30$	**c**	$7000 \div 200$	**d**	$8000 \div 200$
e	$2100 \div 700$	**f**	$9000 \div 60$	**g**	$700 \div 50$	**h**	$3500 \div 70$
i	$3000 \div 500$	**j**	$30\,000 \div 2000$	**k**	$5600 \div 1400$	**l**	$6000 \div 30$

5 Evaluate the following.

a	7.3×10^2	**b**	3.29×10^5	**c**	7.94×10^3	**d**	6.8×10^7
e	$3.46 \div 10^2$	**f**	$5.07 \div 10^4$	**g**	$2.3 \div 10^4$	**h**	$0.89 \div 10^3$

AU 6 You are given that $18 \times 21 = 378$.
Write down the value of:

a 180×210

b $3780 \div 21$

PS 7 Match each calculation to its answer and then write out the calculations in order, starting with the smallest answer.

6000×300 $\quad$ 500×7000 $\quad$ $10\,000 \times 900$ $\quad$ $20 \times 80\,000$

3 500 000 $\quad$ 1 800 000 $\quad$ 1 600 000 $\quad$ 9 000 000

FM 8 The moon is approximately 400 000 km from earth.
If a spaceship takes eight days to reach the moon and return, how far does it travel each day?

8.8 Prime factors, LCM and HCF

HOMEWORK 8I

Example $2^2 \times 3 \times 5 = 4 \times 3 \times 5 = 60$

C

1 Copy and complete the following prime factor trees.

a

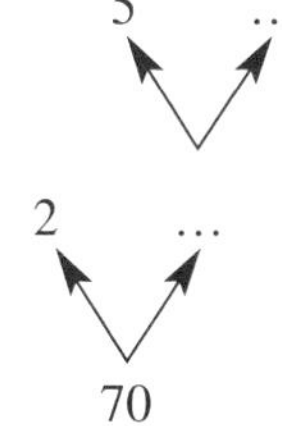

$70 = 2 \times 5 \times \ldots$

b

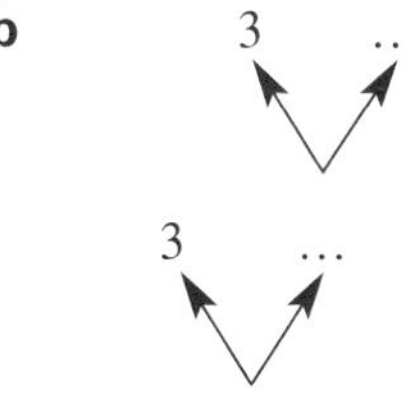

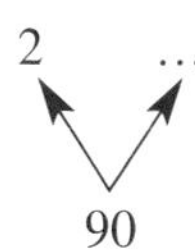

$90 = 2 \times 3 \times 3 \times \ldots$

c

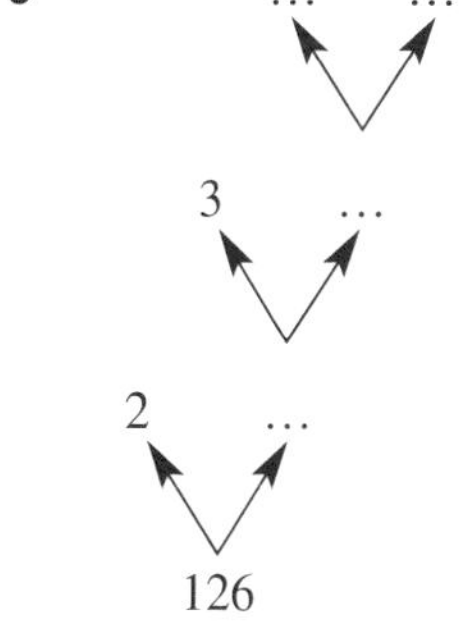

$126 = 2 \times 3 \times \ldots \times \ldots$

2 Write the following as numbers.

a $2^2 \times 3^2$ $\quad$ **b** $2 \times 3 \times 5^2$ $\quad$ **c** $3^2 \times 7$ $\quad$ **d** $2^3 \times 3 \times 5^2$ $\quad$ **e** $3^3 \times 5^2$

3 Write the following numbers as products of their prime factors.

a 24 $\quad$ **b** 36 $\quad$ **c** 75 $\quad$ **d** 84 $\quad$ **e** 99

C

AU **4** **a** Express 45 as a product of prime factors.
b Write your answer to part **a** in index form.
c Use your answer to part **b** to write 90 and 180 as a product of prime factors in index form.

PS **5** $51 = 3 \times 17$
a Write 51^2 as a product of prime factors in index form.
b Write 51^3 as a product of prime factors in index form.

AU **6** The first three odd prime numbers are all factors of 105.
Explain why this means that seven people can share £105 equally so that each receives an exact number of pounds.

HOMEWORK 8J

C

1 Find the LCM of these pairs of numbers.
a 3 and 4 **b** 6 and 8 **c** 9 and 12
d 10 and 12 **e** 14 and 21 **f** 20 and 24

2 Find the HCF of these pairs of numbers.
a 16 and 24 **b** 28 and 35 **c** 24 and 30
d 48 and 60 **e** 28 and 70 **f** 75 and 125

3 For each set of numbers, find **i** the lowest common multiple and **ii** the highest common factor.
a 2, 4 and 6 **b** 4, 6 and 8 **c** 8, 12 and 16 **d** 6, 12 and 15 **e** 20, 25 and 30

FM **4** Nuts are in packs of 12.
Bolts are in packs of 18.
What is the least number of each pack that needs to be bought to have the same number of nuts and bolts?

PS AU **5** The HCF of two numbers is 5.
The LCM of the same two numbers is 150.
What are the numbers?

8.9 Rules for multiplying and dividing powers

HOMEWORK 8K

C

1 Write each of the following as a single power of 7.
a $7^2 \times 7^3$ **b** $7^4 \times 7^5$ **c** 7×7^3 **d** $7^8 \times 7^2$ **e** $7^3 \times 7^4 \times 7^5$

2 Write each of the following as a single power of x.
a $x^2 \times x^3$ **b** $x^4 \times x^5$ **c** $x^6 \times x$ **d** $x^5 \times x^5$ **e** $x^3 \times x^2 \times x^4$

3 Write each of the following as a single power of 4.
a $4^8 \div 4^3$ **b** $4^5 \div 4^2$ **c** $4^7 \div 4^5$ **d** $4^6 \div 4^5$ **e** $4^8 \times 4^4 \div 4^3$

4 Write each of the following as a single power of y.
a $y^5 \div y^2$ **b** $y^8 \div y^3$ **c** $y^{10} \div y$ **d** $y^{12} \div y^4$ **e** $\frac{y^8 \times y^2}{y^3}$

PS **5** **a** Write down the value of $36 \div 36$
b Write $6^2 \div 6^2$ as a single power of 6.
c Use parts **a** and **b** to write down the value 6^0.

AU 6 What happens whenever you divide a number by the same number?

AU 7 $8^a \times 8^b = 8^9$

Write down one pair of possible values for a and b.

Problem-solving Activity

The alternative square root

My square root button has broken.

How can I find a square root using the other buttons?
Here is a method for calculating square roots without using a square root button.

Example $\sqrt{60}$

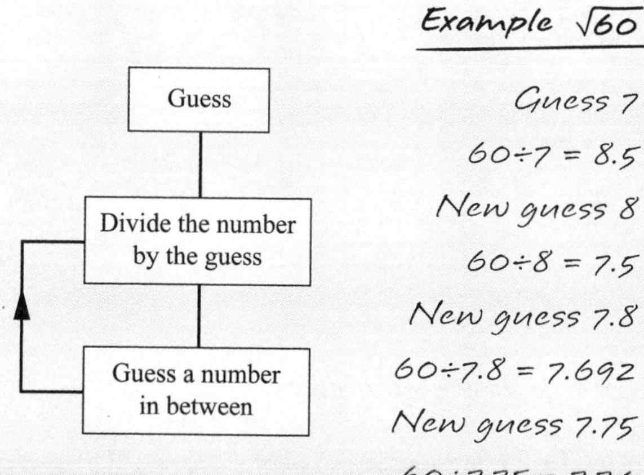

Guess 7

$60 \div 7 = 8.5$

New guess 8

$60 \div 8 = 7.5$

New guess 7.8

$60 \div 7.8 = 7.692$

New guess 7.75

$60 \div 7.75 = 7.74$

Answer is between 7.74 and 7.75

Exact answer 7.7459

Task 1

Use this method to find $\sqrt{1000}$ and then $\sqrt{130}$.
In each case give your answer to 1 decimal place.
Check your answers using the square root button.

Task 2

Choose your own number and use this method to find its square root.

9 Geometry: Symmetry

9.1 Lines of symmetry

HOMEWORK 9A

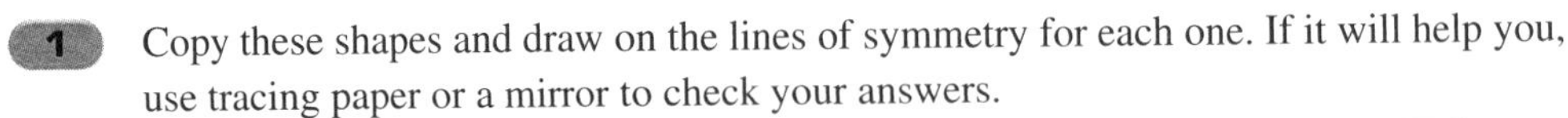

1 Copy these shapes and draw on the lines of symmetry for each one. If it will help you, use tracing paper or a mirror to check your answers.

a 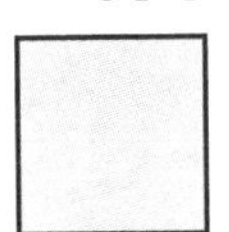 **b** 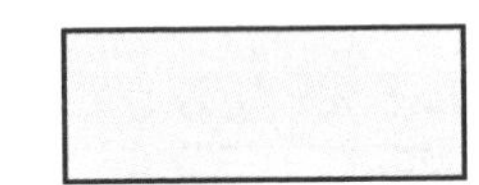 **c** 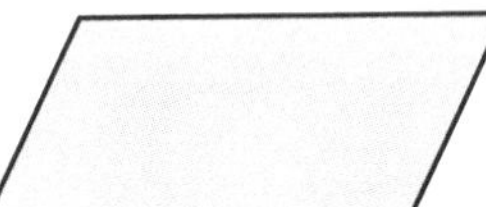

d 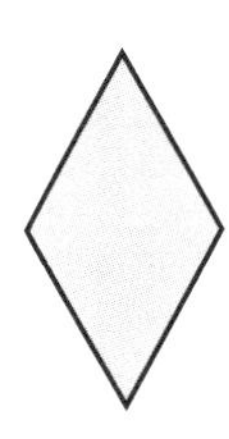 **e** 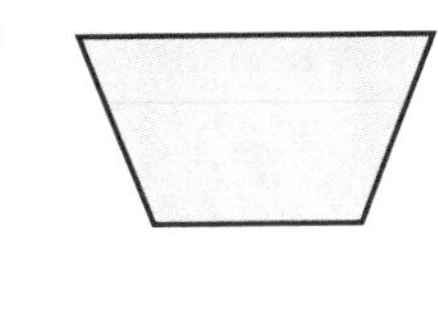

2 Copy this regular hexagon and draw in all the lines of symmetry.

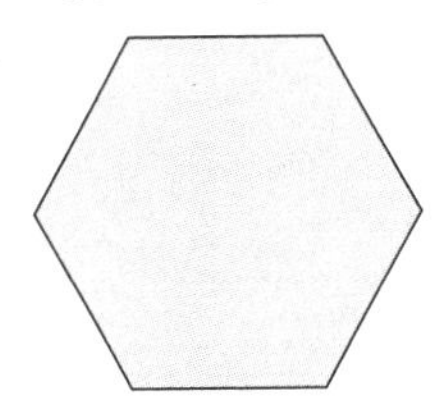

3 Copy these flow chart symbols and draw in all the lines of symmetry for each one.

a 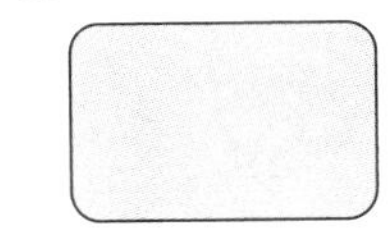 **b** 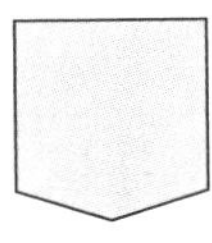 **c** 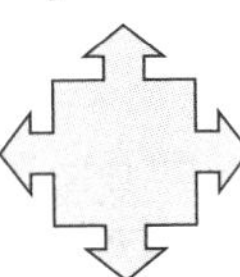

d 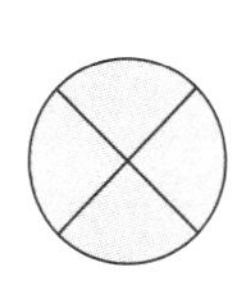 **e** 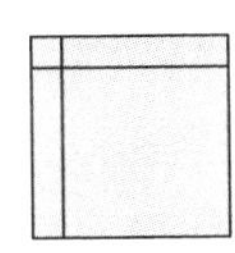

4 Write down the number of lines of symmetry for each of these flags.

a  **b**  **c** 

5 How many lines of symmetry does each of these letters have?

a  **b** E **c**  **d** T **e** 

F

AU 6 Draw three copies of the diagram on the right.

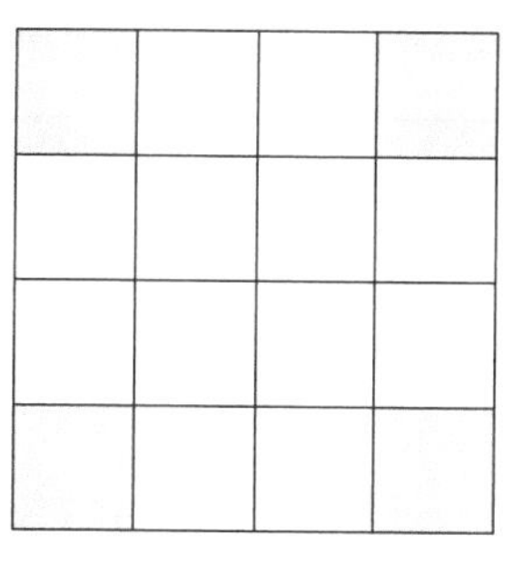

a Shade in two more squares so that the diagram has no lines of symmetry.

b Shade in two more squares so that the diagram has exactly one line of symmetry.

c Shade in two more squares so that the diagram has exactly two lines of symmetry.

7 How many shapes can you find in this garden that have lines of symmetry?

Draw each one and put on the lines of symmetry.

PS 8 Here is a grid and two rectangles.

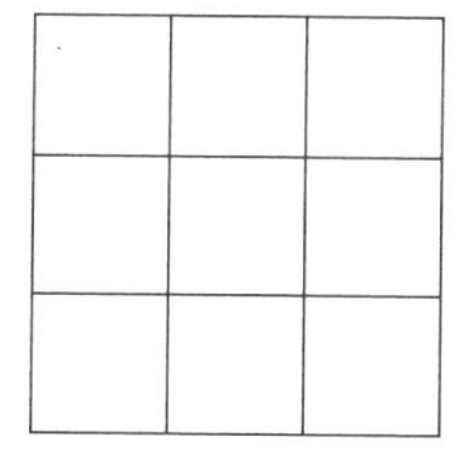

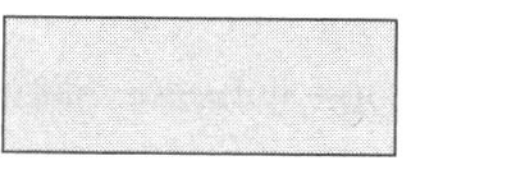

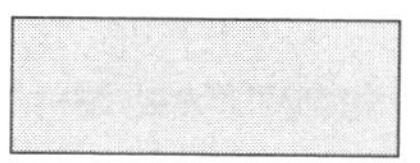

a On a copy of the grid, place the two rectangles so that the grid has one line of symmetry.

b On another copy of the grid, place the two rectangles so that the grid has two lines of symmetry.

AU 9 Which shape is the odd one out?
Give a reason for your answer.

a

b

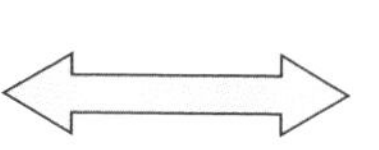

c

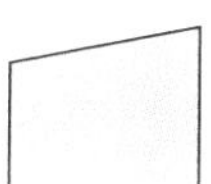

d

9.2 Rotational symmetry

HOMEWORK 9B

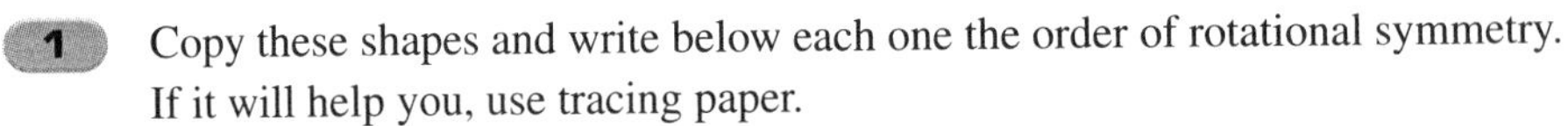

1 Copy these shapes and write below each one the order of rotational symmetry. If it will help you, use tracing paper.

a 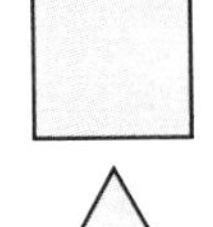 **b** 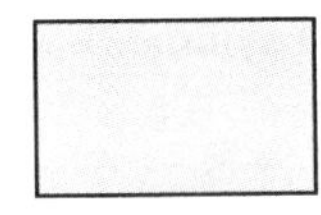 **c** 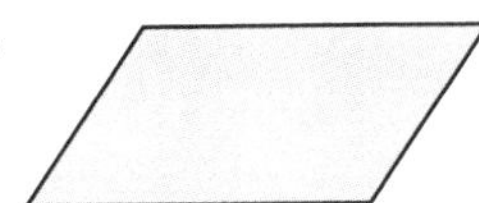

d **e**

2 Write down the order of rotational symmetry for each of these shapes.

a 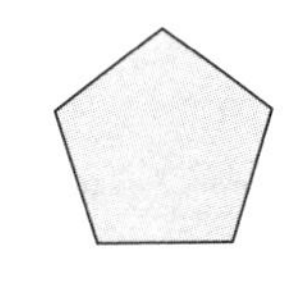 **b** 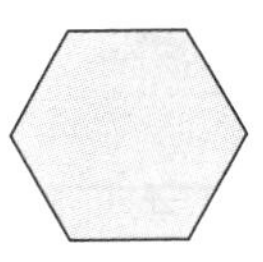 **c** 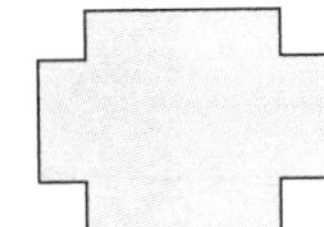

d  **e** 

3 Write down the order of rotational symmetry for each of the symbols.

a 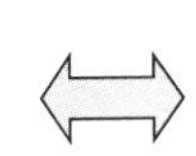 **b** 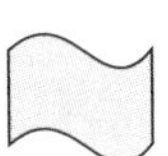 **c** 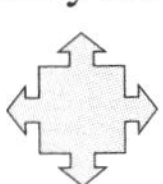 **d**  **e** 

4 The capital letter A fits exactly onto itself only once. So, its order of rotational symmetry is 1. This means that it has no rotational symmetry. Copy these capital letters and write the order of rotational symmetry below each one.

a E **b** H **c** I **d** L **e** N

f Q **g** S **h** Z

AU 5 Draw two copies of the diagram on the right.

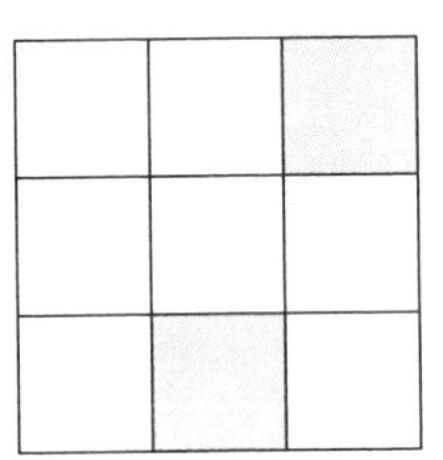

a Shade in two more squares so that the diagram has rotational symmetry of order 2 and no lines of symmetry.

b Shade in two more squares so that the diagram has rotational symmetry of order 1 and exactly 1 line of symmetry.

6 These patterns are taken from old Turkish coins.
What is the order of rotational symmetry for each one?

a

b

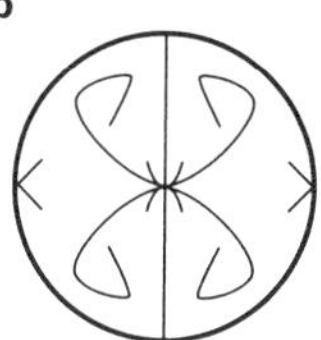

c

d

PS 7 On a copy of this shape, shade in four more squares so that the shape has rotational symmetry of order 2.

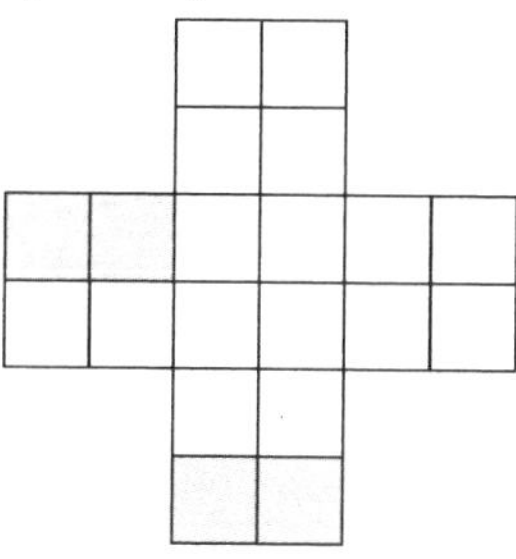

AU 8 Lizzie is drawing shapes that have rotational symmetry of order 3.
Here are some of her examples.

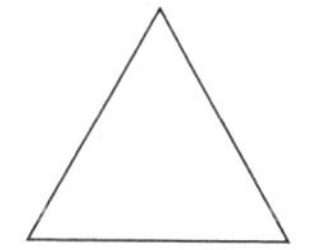

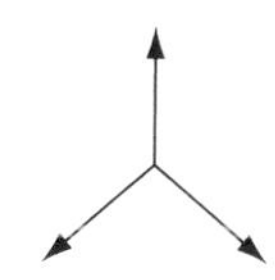

She says that all shapes that have rotational symmetry of order 3 must have three lines of symmetry.
Draw an example to show that she is wrong.

Functional Maths Activity

Symmetry in everyday life

What mathematical questions on symmetry could be asked about this picture?

10 Algebra: Patterns

10.1 Patterns in number

HOMEWORK 10A

Look for the pattern and then write the next two lines. Check your answers with a calculator afterwards.

E

AU 1 $7 \times 11 \times 13 \times 2 = 2002$
$7 \times 11 \times 13 \times 3 = 3003$
$7 \times 11 \times 13 \times 4 = 4004$
$7 \times 11 \times 13 \times 5 = 5005$

AU 2 $3 \times 7 \times 13 \times 37 \times 2 = 20\,202$
$3 \times 7 \times 13 \times 37 \times 3 = 30\,303$
$3 \times 7 \times 13 \times 37 \times 4 = 40\,404$
$3 \times 7 \times 13 \times 37 \times 5 = 50\,505$

AU 3 $3 \times 5 = 4^2 - 1 = 15$
$4 \times 6 = 5^2 - 1 = 24$
$5 \times 7 = 6^2 - 1 = 35$
$6 \times 8 = 7^2 - 1 = 48$

AU 4 $3 \times 7 = 5^2 - 4 = 21$
$4 \times 8 = 6^2 - 4 = 32$
$5 \times 9 = 7^2 - 4 = 45$
$6 \times 10 = 8^2 - 4 = 60$

From your observations on the number patterns above, answer Questions **5** to **9** without using a calculator. Check with a calculator once you have attempted them.

PS 5 $7 \times 11 \times 13 \times 9 =$

PS 6 $3 \times 7 \times 13 \times 37 \times 8 =$

PS 7 $7 \times 11 \times 13 \times 15 =$

PS 8 $3 \times 7 \times 13 \times 37 \times 15 =$

PS 9 $3 \times 7 \times 13 \times 37 \times 99 =$

D

PS 10 This is the calendar for January 2010.

January 2010

Mon	Tue	Wed	Thu	Fri	Sat	Sun
28	29	30	31	1	2	3
4	5	6	7	8	9	10
11	12	13	14	15	16	17
18	19	20	21	22	23	24
25	26	27	28	29	30	31

a Take any 3×3 square, say:

7	8	9
14	15	16
21	22	23

Add up the three numbers in the central column, the central row and the two diagonals. What do you notice?

b What connection is there with the middle number?

c Choose another 3 × 3 square, say:

4	5	6
11	12	13
18	19	20

Without adding any numbers, say what the totals of the three numbers in the central column, the central row and the two diagonals is.

D

10.2 Number sequences

HOMEWORK 10B

D

1 Look at the following number sequences. Write down the next three terms in each and explain how each sequence is found.

a 4, 6, 8, 10, … **b** 3, 6, 9, 12, … **c** 2, 4, 8, 16, …
d 5, 12, 19, 26, … **e** 3, 30, 300, 3000, … **f** 1, 4, 9, 16, …

2 Look carefully at each number sequence below. Find the next two numbers in the sequence and try to explain the pattern.

a 1, 2, 3, 5, 8, 13, 21, … **b** 2, 3, 5, 8, 12, 17, …

3 Look at the sequences below. Find the rule for each sequence and write down its next three terms.

a 7, 14, 28, 56, … **b** 3, 10, 17, 24, 31, … **c** 1, 3, 7, 15, 31, …
d 40, 39, 37, 34, … **e** 3, 6, 11, 18, 27, … **f** 4, 5, 7, 10, 14, 19, …
g 4, 6, 7, 9, 10, 12, … **h** 5, 8, 11, 14, 17, … **i** 5, 7, 10, 14, 19, 25, …
j 10, 9, 7, 4, … **k** 200, 40, 8, 1.6, … **l** 3, 1.5, 0.75, 0.375, …

C

FM 4 A well known rhyme to pick someone from a group is:

Eeny, meeny, miney, moe,
Catch a baby by the toe.
If he cries, let him go,
Eeny, meeny, miney, moe.

Each time a word is spoken, a different person is pointed at.
The person pointed at on the last 'moe' is picked or is 'out'.
Imagine 10 people – Alexander, Briony, Chris, David, Ellie, Fran, Greta, Hermione, Isabel and Jack – are standing in a circle.
Another person, Xavier, stands in the middle and starts by pointing at Alexander with the first 'Eeny'.
Check that the person that the last 'moe' lands on is Jack.
Jack is out and there are now only nine people left in the circle. The rhyme is repeated, starting with the next person, who happens to be Alexander again. Once again, the person that the last 'moe' lands on is out.
The process carries on until only one person is left.

a Who is the last person left?

b What order were they eliminated in?

PS 5 Two sequences are:

5, 11, 17, 23, 26, 32, 38, ...
1, 4, 7, 10, 13, 16, 19, ...

Will the two sequences ever have any terms in common? Explain your answer.

C

AU 6 Two sequences are
100, 96, 92, 88, 84, ...
2, 8, 14, 20, 26, ...
Find all the terms that the sequences have in common.

10.3 The *n*th term of a sequence

HOMEWORK 10C

D

1 Use each of the following rules to write down the first five terms of a sequence.

a $3n + 1$ for $n = 1, 2, 3, 4, 5$
b $2n - 1$ for $n = 1, 2, 3, 4, 5$
c $4n + 2$ for $n = 1, 2, 3, 4, 5$
d $2n^2$ for $n = 1, 2, 3, 4, 5$
e $n^2 - 1$ for $n = 1, 2, 3, 4, 5$

2 Write down the first five terms of the sequence which has its *n*th term as:

a $n + 2$ **b** $4n - 1$ **c** $4n - 3$ **d** $n^2 + 1$ **e** $2n^2 + 1$

3 The *n*th term of a sequence is $\frac{n}{2n - 1}$

The first term of this sequence is $\frac{1}{2 \times 1 - 1}$

Work out the first five terms of this sequence as fractions.

C

FM 4 A physiotherapist uses this formula for charging for a series of *n* sessions if paid for in advance.
For $n \leqslant 5$, cost will be £$(35n + 20)$
For $6 \leqslant n \leqslant 10$, cost will be £$(35n + 10)$
For $n \geqslant 11$, cost will be £$35n$

a How much will the physiotherapist charge for 8 sessions booked in advance?
b How much will the physiotherapist charge for 14 sessions booked in advance?
c One client paid £220 in advance for a series of sessions.
How many sessions did she book?
d A runner has a leg injury and is not sure how many sessions it will take to cure.
The runner books 4 sessions in advance, and after the sessions starts to run in races again. The leg injury returns and he has to book another 3 sessions before he is finally cured.
How much more did it cost him than if he had booked 7 sessions in advance?

AU 5 The formula for working out a series of fractions is $\frac{n + 1}{2n + 1}$
Show that in the first 8 terms only one of the fractions is a terminating decimal.

HINTS AND TIPS

If you set this up on a spreadsheet, find the relationship between the denominators of the terms that give terminating decimals in this series.

AU 6 The nth term of a sequence is $106 - 4n$.
The nth term of another sequence is $6n - 4$.
These two series have several terms in common, for example, 98, 86, 74, ... but only one term that is common and has the same position in the sequence.
Without writing out the sequences, show how you can tell, using the expressions for the nth term, that this is the 11th term.

HOMEWORK 10D

1 Find the nth term in each of these linear sequences.

a 5, 7, 9, 11, 13 … **b** 6, 10, 14, 18, 22, … **c** 6, 11, 16, 21, 26, …
d 3, 9, 15, 21, 27, … **e** 4, 7, 10, 13, 16, … **f** 3, 10, 17, 24, 31, …

2 Find the 50th term in each of these linear sequences.

a 3, 5, 7, 9, 11, … **b** 5, 9, 13, 17, 21, … **c** 8, 13, 18, 23, 28, …
d 2, 8, 14, 20, 26, … **e** 5, 8, 11, 14, 17, … **f** 2, 9, 16, 23, 30, …

3 For each sequence **a** to **f**, find:

i the nth term **ii** the 100th term **iii** the term closest to 100.

a 5, 12, 19, 26, 33, ... **b** 9, 11, 13, 15, 17, ...
c 2, 7, 12, 17, 22, ... **d** 2, 6, 10, 14, 18, ...
e 5, 13, 21, 29, 37, ... **f** 6, 7, 8, 9, 10, ...

AU 4 The cube numbers are 1, 8, 27, 64, 125, ...
The nth term of this sequence is given by n^3

a Work out the 10th cube number.
b Write down the nth terms of these sequences.
i 2, 9, 28, 65, 126, ... **ii** 2, 8, 54, 128, 250, ...
iii 0.5, 4, 13.5, 32, 62.5, ...

FM 5 This chart is used by a taxi firm for the charges for journeys of k kilometres.

k	1	2	3	4	5	6	7	8	9	10
Charge (£)	4.50	6.50	8.50	10.50	12.50	15.00	17.00	19.00	21.00	23.00
k	11	12	13	14	15	16	17	18	19	20
Charge (£)	26.00	28.00	30.00	32.00	34.00	37.00	39.00	41.00	43.00	45.00

a Using the charges for 1 to 5 kilometres, work out an expression for the kth term.
b Using the charges for 6 to 10 kilometres, work out an expression for the kth term.
c Using the charges for 10 to 15 kilometres, work out an expression for the kth term.
d Using the charges for 16 to 20 kilometres, work out an expression for the kth term.
e What is the basic charge per kilometre?

PS 6 A series of fractions is $\frac{3}{7}, \frac{5}{10}, \frac{7}{13}, \frac{9}{16}, \frac{11}{19}, \ldots$

a Write down an expression for the nth term of the numerators.
b Write down an expression for the nth term of the denominators.
c **i** Work out the fraction when $n = 1000$.
ii Give the answer as a decimal.
d Will the terms of the series ever be greater than $\frac{2}{3}$?
Explain your answer.

10.4 Special sequences and algebra

HOMEWORK 14E

1 The powers of 3 are $3^1, 3^2, 3^3, 3^4, 3^5, \ldots$.
This give the sequence 3, 9, 27, 81, 243, ….

a Continue the sequence for another 3 terms.

b The nth term is given by 3^n.
Give the nth terms of each of these sequences.

i 2, 8, 26, 80, 242, ….

ii 6, 18, 54, 162, 486

PS 2 p is an odd number and q is an even number. State whether the following are odd or even.

a $p + 5$ **b** $q - 3$ **c** $2p$

d q^2 **e** pq **f** $2(p + q)$

g $p^2 + q$ **h** $q(p + q)$

AU 3 Write down the next two lines of this number pattern.

$0 + 1 = 1 = 1^2$
$1 + 3 = 4 = 2^2$
$3 + 6 = 9 = 3^2$
$6 + 10 = 16 = 4^2$

PS 4 P is a prime number, Q is an odd number and R is an even number.
State if the following are always odd (O), always even (E) or could be either (C).

a P + 2 **b** P + Q **c** PR + Q2 **d** $(P + Q)(P + R)$

10.5 General rules from given patterns

HOMEWORK 10F

FM 1 A conference centre has tables that can each sit three people. When put together, the tables can seat people as shown.

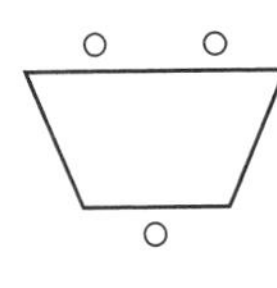
1

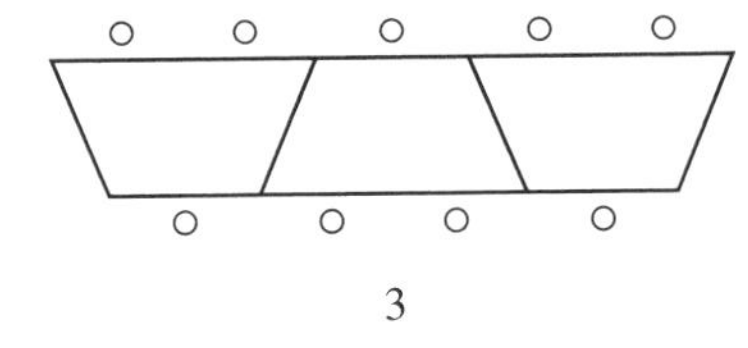
2 3

a How many people could be seated at four tables?

b How many people could be seated at n tables put together in this way?

c A conference had 50 people who wished to use the tables in this way. How many tables would they need?

2 A pattern of shapes is built up from matchsticks as shown.

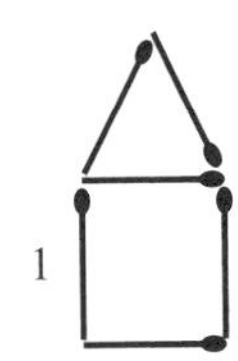
1

2

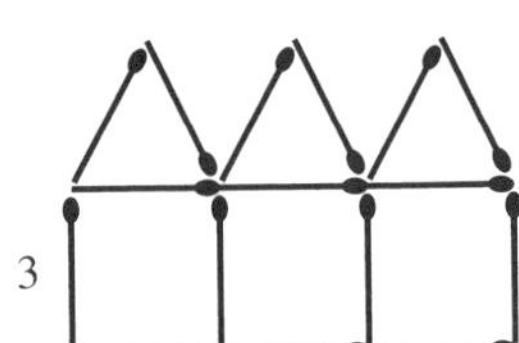
3

a Draw the fourth diagram.

b How many matchsticks are in the nth diagram?

c How many matchsticks are in the 25th diagram?

d With 200 matchsticks, which is the biggest diagram that could be made?

3 A pattern of hexagons is built up from matchsticks.

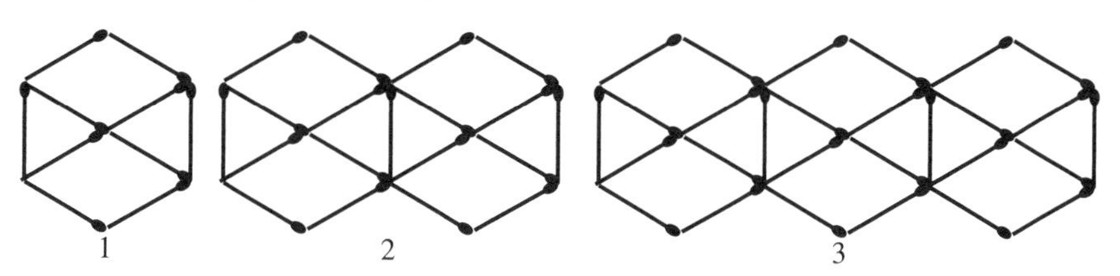

a Draw the fourth set of hexagons in this pattern.

b How many matchsticks are needed for the nth set of hexagons?

c How many matchsticks are needed to make the 60th set of hexagons?

d If there are only 100 matchsticks, that is the largest set of hexagons that could be made?

PS 4 **a** Draw an equilateral triangle with each side 9 cm.
The perimeter will be 27 cm.

b Draw another equilateral triangle of side 3 cm on each edge.
Work out the perimeter of the new shape.

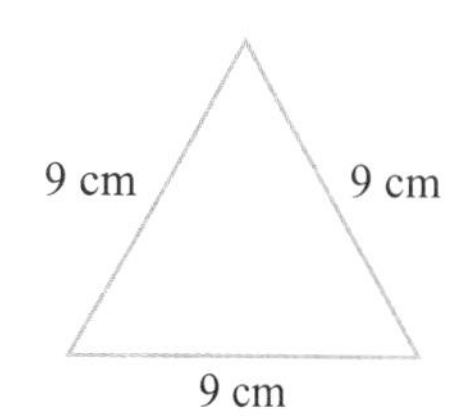

c Draw another equilateral triangle of side 1 cm on each of the remaining edges.
Work out the perimeter of the new shape.

d The next step would be to draw a triangle of side $\frac{1}{3}$ cm on each remaining edge, but this will be difficult to draw.

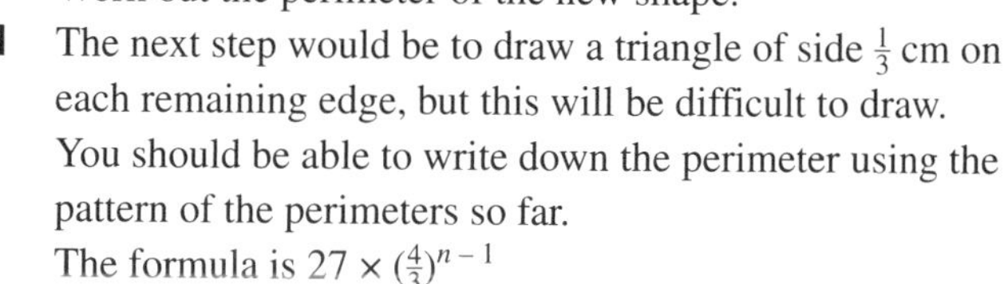

You should be able to write down the perimeter using the pattern of the perimeters so far.
The formula is $27 \times (\frac{4}{3})^{n-1}$

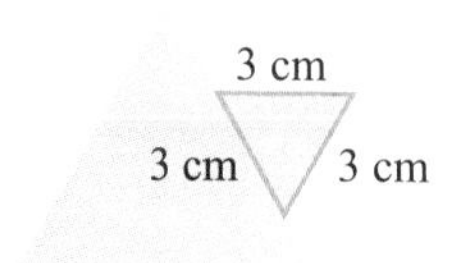

You will need a calculator with a power button (^).
Work out 27 × (4 ÷ 3) ^ 0, which should equal 27.
Then work out 27 × (4 ÷ 3) ^ 1, which should equal your answer to the perimeter in part **b**.
Use the formula to work out the perimeter of the next drawing when $n = 4$

e Work out the perimeter when $n = 100$.
If we kept on drawing triangles, the perimeter would become infinite.
This is an example of a shape that has a finite area surrounded by an infinite perimeter.

FM 5 For a display of grapefruits, a supermarket manager stacks them in layers, each of which is a triangle.
These are the first four layers.

 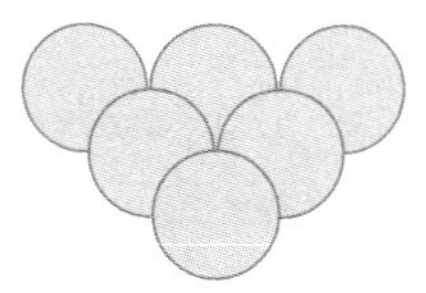  

a If the display is four layers deep, how many grapefruits will be in the display?

b The manager tells his staff that there should not be any more than eight layers, as the fruit will get squashed otherwise.
What is the most grapefruits that could be stacked?

AU 6 Harry is building three different patterns with counters.
He builds the patterns in steps.

	Step 1	Step 2	Step 3	Step 4
Pattern 1	● ● ●	●● ●● ●●	●●● ●●● ●●●	●●●● ●●●● ●●●●
Pattern 2	● ●	●● ●●	●●● ●●●	●●●● ●●●●
Pattern 3	●	●●	●●●	●●●●

Harry has a packet that contains 1000 counters.
Which step will Harry get to before he runs out of counters?

Problem-solving Activity

Pascal's triangle

Pascal's triangle has many uses in mathematics.

Each row starts and ends with 1. The other numbers are formed by adding the two values above them.

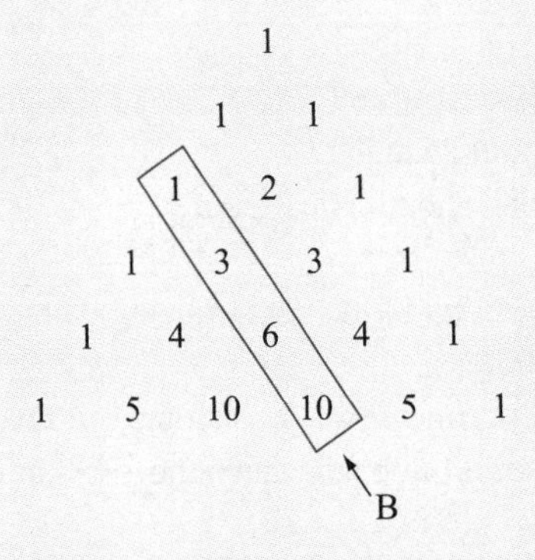

1. Continue Pascal's triangle for another three rows.
2. Describe any patterns or special sequences you can see in Pascal's triangle.
3. What is the special name given to the series of numbers down the diagonal marked B?
4. Add each row, e.g. 1 = 1, 1 + 1 = 2, 1 + 2 + 1 = 4.
   Explain how the series formed is building up.
5. Use the internet to find out about Blaise Pascal and his triangle.
   How can it be used in statistics?

11 Number: Percentages

11.1 Rational numbers and reciprocals

HOMEWORK 11A 

C

1 Work out each of these fractions as a decimal. Give them as terminating decimals or recurring decimals as appropriate.

a $\frac{3}{4}$ **b** $\frac{1}{15}$ **c** $\frac{1}{25}$ **d** $\frac{1}{11}$ **e** $\frac{1}{20}$

2 There are several patterns to be found in recurring decimals. For example,

$\frac{1}{13} = 0.076923076923076923076923 0\ldots$, $\frac{2}{13} = 0.153846153846153846153846\ldots$,

$\frac{3}{13} = 0.230769230769230769230769\ldots$ and so on.

a Write down the decimals for $\frac{4}{13}, \frac{5}{13}, \frac{6}{13}, \frac{7}{13}, \frac{8}{13}, \frac{9}{13}, \frac{10}{13}, \frac{11}{13}, \frac{12}{13}$ to 24 decimal places.

b What do you notice?

3 Write each of these fractions as a decimal. Use this to write the list in order of size, smallest first.

$\frac{2}{9}$ $\frac{1}{5}$ $\frac{23}{100}$ $\frac{2}{7}$ $\frac{3}{11}$

4 Convert each of these terminating decimals to a fraction in its simplest form.

a 0.57 **b** 0.275 **c** 0.85 **d** 0.06 **e** 3.65

5 Use a calculator to work out the reciprocal of each of the following.

a 4 **b** 8 **c** 32 **d** 40 **e** 100

6 Write down the reciprocal of each of the following fractions.

a $\frac{2}{3}$ **b** $\frac{5}{8}$ **c** $\frac{9}{10}$ **d** $\frac{7}{12}$ **e** $\frac{17}{20}$

AU 7 Explain why the reciprocal of 1 is 1.

PS 8
a Work out the reciprocal of the reciprocal of 4.
b Work out the reciprocal of the reciprocal of 5.
c What do you notice

11.2 Increasing or decreasing quantities by a percentage

HOMEWORK 11B 

Example Increase £6 by 5%.

Method 1 Find 5% of £6: $(5 \div 100) \times 6 = £0.30$
Add the £0.30 to the original amount: $£6 + £0.30 = £6.30$

Method 2 Using a multiplier: $1.05 \times 6 = £6.30$

D

1 Increase each of the following by the given percentage. (Use any method you like.)

a £80 by 5% **b** £150 by 10% **c** 800 m by 15% **d** 320 kg by 25%
e £42 by 30% **f** £24 by 65% **g** 120 cm by 18% **h** £32 by 46%
i 550g by 85% **j** £72 by 72%

D

2 Mr Kent, who was on a salary of £32 500, was given a pay rise of 4%. What is his new salary?

FM **3** Copy and complete this electricity bill.

	Total charges
Fixed charges	£13.00
840 units @ 6.45 p per unit	
1720 units @ 2.45 p per unit	
Total charges	
VAT @ 8%	
Total to pay	

4 A bank pays 8% simple interest on the money that each saver keeps in a savings account for a year. Miss Pettica puts £2000 in this account for three years. How much will she have in her account after:

a 1 year **b** 2 years **c** 3 years?

FM **5** VAT (Value Added Tax) is a tax that the Government adds to the price of goods sold. At the moment it is 17.5% on all goods. Mrs Dow purchased these items from a gift catalogue, after VAT of 17.5% has been added.

Gift	*Pre-VAT price*
Travel alarm clock	£18.00
Ladies' purse wallet	£15.20
Pet's luxury towel	£12.80
Silver-plated bookmark	£6.40

She estimated that the total cost would be about £60. Was this a good estimate? Show how you decide.

PS FM **6** A dining table costs £300 before the VAT is added.
If the rate of VAT goes up from 15% to 20%, by how much will the cost of the dining table increase?

HOMEWORK 11C

Example Decrease £6 by 5%.

Method 1 Find 5% of £6: $(5 \div 100) \times 6 = £0.30$
Subtract the £0.30 from the original amount: $£6 - £0.30 = £5.70$

Method 2 Using a multiplier: $0.95 \times 6 = £5.70$

D

1 Decrease each of the following by the given percentage. (Use any method you like.)

a £20 by 10% **b** £150 by 20% **c** 90 kg by 30% **d** 500 m by 12%
e £260 by 5% **f** 80 cm by 25% **g** 400 g by 42% **h** £425 by 23%
i 48 kg by 75% **j** £63 by 37%

D

FM **2** Mrs Denghali buys a new car from a garage for £8400. The garage owner tells her that the value of the car will decrease by 24% after one year. What will be the value of the car after one year?

3 The population of a village in 2006 was 2400. In 2010 the population had decreased by 12%. What was the population of the village in 2010?

FM **4** A Travel Agent is offering a 15% discount on holidays. How much will the advertised holiday now cost?

NEW YORK FOR A WEEK
£540

5 

Matt has £160 from Christmas presents. Can he afford to buy a shirt that normally costs £30, a suit that normally costs £130, and a pair of shoes that normally cost £42?

C

PS **6** A shop increases all its prices by 10%.
One month later it advertises 10% off all marked prices.
Are the goods cheaper, the same or more expensive than before the price increase?
Show how you work out your answer.

11.3 Expressing one quantity as a percentage of another quantity

HOMEWORK 11D

Example Express £6 as a percentage of £40.

Set up the fraction $\frac{6}{40}$ and multiply it by 100. $6 \div 40 = 15\%$.

D

1 Express each of the following as a percentage. Give your answers to one decimal place where necessary.

a £8 of £40 **b** 20 kg of 80 kg **c** 5 m of 50 m
d £15 of £20 **e** 400 g of 500 g **f** 23 cm of 50 cm
g £12 of £36 **h** 18 minutes of 1 hour **i** £27 of £40
j 5 days of 3 weeks

2 What percentage of these shapes is shaded?

a 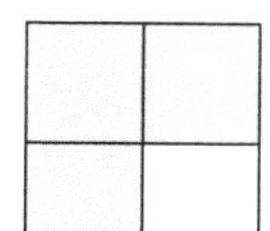 **b** 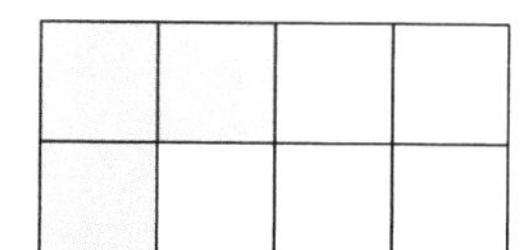

3 In a class of 30 pupils, 18 are girls.
a What percentage of the class are girls?
b What percentage of the class are boys?

D

4 The area of a farm is 820 hectares. The farmer uses 240 hectares for pasture.
What percentage of the farm land is used for pasture? Give your answer to one decimal place.

C

FM 5 Here are some retail and wholesale prices:

	Item	*Retail price (Selling price)*	*Wholesale price (Price the shop paid)*
a	Micro Hi-Fi System	£250	£150
b	CD Radio Cassette	£90	£60
c	MiniDisc Player	£44.99	£30
d	Cordless Headphones	£29.99	£18

A shopkeeper wants to make over 40% profit on each item. Does he succeed at these prices?

AU 6 Paul and Val take the same tests. Both tests are worth the same number of marks.
Here are their results.

	Test A	**Test B**
Paul	30	40
Val	28	39

Whose result has the greater percentage increase from test A to test B?
Show your working.

HOMEWORK 11E

F

1 Copy and complete the table

	Fraction	**Decimal**	**Percentage**
a	1/4		
b		0.4	
c			15%

E

2 Work out these amounts.
a 15% of £42
b 12% of 300 kg
c 35% of 240 ml

D

3 What percentage is:
a 36 out of 50
b 17 out of 25
c 60 out of 200

4 What is the result if:
a 180 is increased by 25%
b 4200 is decreased by 7%

5 **a** A window-cleaner increases his fee from £12 to £15 per house. What is the percentage increase in his fee?
b The number of houses on his round increases from 40 to 48. What is the percentage increase in the number of houses he cleans windows for?
c For cleaning the windows of a bungalow, he offers a 30% discount on his new fee. What does he charge a bungalow-owner?

6 **a** A new computer costs £800 at full price. In it's winter sale, a computer shop offers a 20% discount. What is the sale price of the computer?

b The shop is also selling some computer software at a 15% discount. Before the sale it cost £120. Patricia decides to buy both the computer and the software in the sale. She has been saving £75 per month for a year. Does she have enough money?

AU PS **7** A group of mothers agreed to compare the weights of their newly born babies. The mothers said their babies had a mean weight of 4 kg when they were first born. The mothers said they would compare the weights one month later to see the average amount of weight they gained. When the babies were one month old the mothers said their babies had gained an average of 25% in weight.

a What was the average weight after the month?

b One of the mothers realised she had misread the scale and her baby was 0.5 kg heavier than she thought. Which of these statements is true?

i The mean weight gain will have increased by more than 25%.

ii The mean weight gain will have stayed at 25%.

iii The mean weight gain will have increased by less than 25%.

iv There is not enough information to answer the question.

C

Functional Maths Activity

The cost of going to work

Miss Jones

- Miss Jones is 23 years old and lives in Bramley.
- She works 20 miles from home, in Aston.
- She is a manager in a small company and earns £18 000 per year.
- She works from Monday to Friday each week.
- She has four weeks holiday per year.
- She always takes two of these holiday weeks in July every year.
- She travels to work by train each day using a monthly ticket.
- She has a 16–25 Railcard.
- In July she buys weekly tickets.
- The journey takes 45 minutes each way.
- She uses the local sandwich shop for lunch each day.

Rail Fares
Bramley to Aston
7-Day ticket £56.40
Monthly ticket £217.35
3-month ticket £652.05

16–25 Railcard — £26 for a whole year

Aged 16–25 or a full-time student aged 26 or over?

Save 1/3 on most rail fares throughout Great Britain

16–25 Railcard discounts now apply to *all*
Standard and First Class Advance fares.

Sandwich shop
Small sandwiches ~ £2.50
Large Sandwiches ~ £3.30
Pay weekly for your sandwiches and get Friday free!

Mr Smith

- Mr Smith is 45 years old and lives in Sunnyside.
- He works 10 miles from home, in Todwick.
- He is a maintenance worker in the same small company and earns £12000 per year.
- He works from Monday to Saturday each week.
- He has six weeks holiday per year.
- He travels to work by bus each day, using a daily return ticket.
- The journey takes 30 minutes each way.
- He has lunch in the works canteen each day.
- He always has the set meal, plus a drink and two portions of extra vegetables.

Bus fares
Todwick to Aston
Single £2.35
Return £3.20

WORKS CANTEEN

SET MEAL --- £4
DRINKS --- 75P EACH
EXTRA VEGETABLES 50P PER PORTION

Functional Maths Activity (continued)

Task 1

Answer the following questions about Miss Jones.

1. How many weeks does she work in a year?
2. How much is she paid each month?
3. How much does she pay for a monthly rail ticket?
4. Why does she buy weekly tickets in July?
5. What percentage of her salary does she spend on travel to work, including the cost of the Railcard, in one year?
6. How much would she pay at the sandwich shop if she pays weekly for small sandwiches? What percentage is this of her salary per month?
7. What percentage of the cost of the sandwiches is she saving?
8. How much more per week would it cost her if she had large sandwiches?
9. What percentage of her monthly salary is the answer to question 8?
10. One-third of her salary is spent on taxes. How much does she have left after tax? What percentage is this?
11. Rework questions 6 and 7 according to your answer in question 10.

Task 2

Use your answers to **Task 1** to help you to work out how much money Miss Jones has left after deducting taxes, travelling and meal costs from her salary.
Give your answer as a monthly amount. What percentage is this of her total salary?

Task 3

Work the time that Miss Jones spends travelling to and from work each year. What percentage is this?

Task 4

Work out how much money Mr Smith has left after deducting taxes, travelling and meal costs from his salary per year. What percentage is this of his annual salary?

Task 5

- Imagine that you live 15 miles from work.
- Decide costs for travelling to work by train or bus.
- Decide what you will eat at lunchtime.
- Decide what your salary will be.
- If the salary is low, deduct one-third of the salary for taxes.
- If the salary is high, the ratio of tax to remaining pay is 2 : 1.
- Work out how much money you will have left after deducting taxes, travelling and meal costs from your salary. What percentage is this?

12 Algebra: Further algebra

12.1 Solving linear equations

HOMEWORK 12A

F

E

1 Solve the following equations.

a $x + 2 = 8$ **b** $y - 4 = 3$ **c** $s + 7 = 10$ **d** $t - 7 = 4$

e $3p = 12$ **f** $5q = 15$ **g** $\frac{k}{2} = 4$ **h** $4n = 20$

i $\frac{a}{3} = 2$ **j** $b + 1 = 2$ **k** $c - 7 = 7$ **l** $\frac{d}{5} = 1$

AU 2 The solution to the equation $\frac{x}{4} = 6$ is $x = 24$.

Write down two **different** equations for which the solution is 24.

AU 3 Here are three equations.

A: $\frac{x}{3} = 6$ B: $\frac{36}{x} = 2$ C: $\frac{x}{2} = 36$

a Give one similarity between A and B.
b Give one similarity between A and C.
c Give one similarity between B and C.

4 Set up an equation to represent the following. Use x for the variable.
My mother is twice as old as me. She is 38 years old. How old am I?

5 Set up an equation to represent the following. Use y for the variable.
10 litres of petrol cost £9.50. How much is one litre?

HOMEWORK 12B

Example Solve $3x - 4 = 11$ using an inverse flow diagram.

The flow diagram for the equation is:

$x \longrightarrow \boxed{\times 3} \longrightarrow \boxed{-4} \longrightarrow 11$

Inverse flow diagram:

$x \longleftarrow \boxed{\div 3} \longleftarrow \boxed{+4} \longleftarrow 11$

Put through the value on the right-hand side:

$5 \longleftarrow \boxed{\div 3} \longleftarrow \boxed{+4} \longleftarrow 11$

The answer is $x = 5$

Checking the answer gives $3 \times 5 - 4 = 11$ which is correct.

E

1 Solve each of the following equations using inverse flow diagrams. Do not forget to check that each answer works in the original equation.

a $2x + 5 = 13$ **b** $3x - 2 = 4$ **c** $2x - 7 = 3$ **d** $3y - 9 = 9$

e $5a + 1 = 11$ **f** $4x + 5 = 21$ **g** $6y + 6 = 24$ **h** $5x + 4 = 9$

i $8x - 10 = 30$ **j** $\frac{x}{2} + 1 = 4$ **k** $\frac{a}{2} - 2 = 3$ **l** $\frac{c}{3} + 2 = 8$

m $\frac{x}{3} - 3 = 1$ **n** $\frac{m}{3} - 1 = 2$ **o** $\frac{z}{5} + 6 = 10$

D

AU 2 The reverse flow diagram shows the solution to an equation.

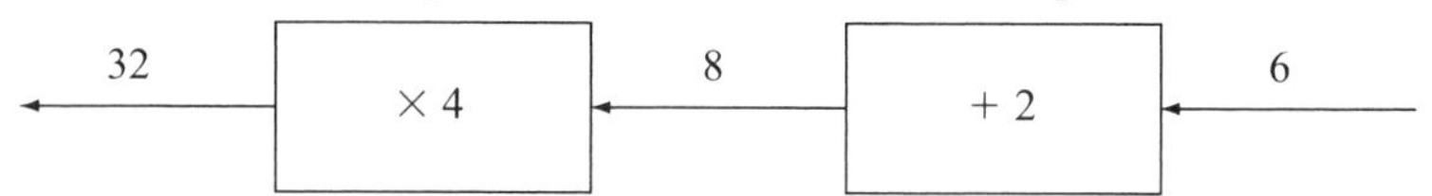

What is the equation?

PS 3 The diagram shows a two-step number machine.

Find a value for the input that gives the same value for the output.

HOMEWORK 12C

Example Solve the equation $3x - 5 = 16$ by 'doing the same to both sides'.

$3x - 5 = 16$ Add 5 to both sides

$3x - 5 + 5 = 16 + 5$

$3x = 21$ Divide both sides by 3

$\frac{3x}{3} = \frac{21}{3}$

$x = 7$

E

1 Solve each of the following equations by 'doing the same to both sides'. Do not forget to check that each answer works in the original equation.

a $x + 5 = 6$ **b** $y - 3 = 4$ **c** $x + 5 = 3$ **d** $2y + 4 = 12$

e $3t + 5 = 20$ **f** $2x - 4 = 12$ **g** $6b + 3 = 21$ **h** $4x + 1 = 5$

i $2m - 3 = 4$ **j** $\frac{x}{2} - 5 = 2$ **k** $\frac{a}{3} + 3 = 6$ **l** $\frac{z}{5} - 1 = 1$

D

AU 2 The solution of the equation $\frac{x}{3} + 8 = 16$ is $x = 24$.

Make up **two** more **different** equations of the form $\frac{x}{a} + b = c$ where a, b and c are positive whole numbers for which the answer is also 24.

D

PS **3** Two students solved the equation $\frac{x}{2} + 3 = 5$ in two different ways:

Student 1	**Student 2**
$\frac{x}{2} + 3 = 5$	$\frac{x}{2} + 3 = 5$
$\frac{x}{2} + 3 - 3 = 5 - 3$	$\frac{x}{2} + 3 + 3 = 5 + 3$
$\frac{x}{2} = 2$	$\frac{x}{2} = 8$
$\frac{x}{2} \times 2 = 2 \times 2$	$\frac{x}{2} \div 2 = 8 \div 2$
$x = 4$	$x = 4$

a Which student used the correct method?

b Explain the mistakes the other student made.

HOMEWORK 12D

Example Solve $4x + 3 = 23$.

Subtract 3 to give $4x = 23 - 3 = 20$

Now divide both sides by 4 to give $x = 20 \div 4 = 5$

The solution is $x = 5$

E

1 Solve each of the following equations. Do not forget to check that each answer works in the original equation.

a $2x + 1 = 7$ **b** $2t + 5 = 13$ **c** $3x + 5 = 17$ **d** $4y + 7 = 27$

e $2x - 8 = 12$ **f** $5t - 3 = 27$ **g** $\frac{x}{2} + 3 = 6$ **h** $\frac{p}{3} + 2 = 3$

i $\frac{x}{2} - 3 = 5$ **j** $8 - x = 2$ **k** $13 - 2k = 3$ **l** $6 - 3z = 0$

D

2 Solve each of these equations.

a $\frac{x + 2}{3} = 4$ **b** $\frac{y - 4}{5} = 2$ **c** $\frac{z + 4}{8} = 5$

AU **3** A teacher reads out the following to her class:
'I am thinking of a number. I multiply it by 2 and then subtract 3. The answer is 12. What number did I think of to start with?'

a What was the number the teacher thought of?

b Ben misunderstood the instructions and got the operations the wrong way round. What number did Ben think the teacher started with?

PS **4** Six boxes of apples each holding A apples are delivered to a supermarket.
18 of the apples are found to be bad and thrown away.
The rest are packed into 45 trays with six apples in each before being put on the shelves.
How many apples, A, are in each box?

12.2 Setting up equations

HOMEWORK 12E

Set up an equation to represent each situation described below. Then solve the equation. Do not forget to check your answer to each question.

1 A teacher asks her class to think of a number and subtract 6 from it.

a What was Alan's final answer?

b What was Vikram's original answer?

AU 2 A class of 24 students had a collection to buy some chocolates for their teacher's birthday. Each student gave p pence and the teaching assistant gave a pound. The chocolates cost £10.60.

a Which of the following equations represents this situation?

$24p + 1 = 10.6$ $\quad$ $24p + 100 = 10.6$ $\quad$ $24p + 100 = 1060$

b How much did each student contribute?

FM 3 This diagram shows the way that water flows through a pipe network.

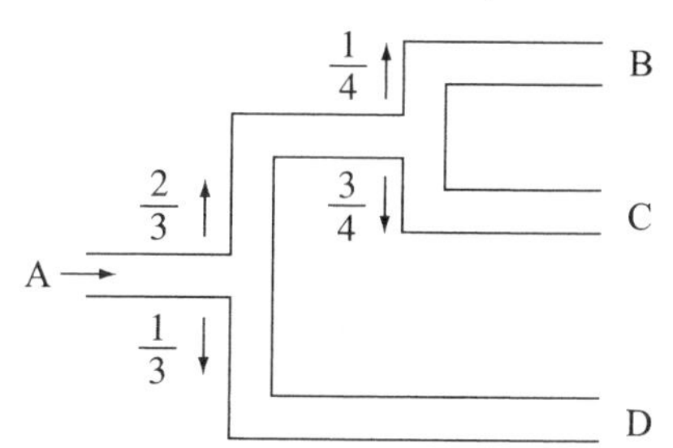

Water enters at A, and at each junction the fractions show the proportion of water that take each route.

a 9000 gallons enter at A. How many gallons come out of each of exits B, C and D?

b If 1200 gallons exit at B, how many gallons entered at A?

c If 4800 gallons exit at D, how many gallons exit at B?

FM 4 Martin bought 12 bottles of pop. When he got to the till, he used a £2 coupon as part payment. His final bill was £7.

a Set this problem up as an equation using p as the cost of one bottle of pop.

b Solve the equation to work out the cost of one bottle of pop.

PS 5 A rectangular room is 4 metres longer than it is wide. The perimeter is 28 metres. It cost Mr Plush £607.50 to carpet the room. How much is the carpet per square metre?

D

C

FM **6** Books cost twice as much as magazines.
Kerry buys the same number of books and magazines and pays £22.50
Derek buy one book and two magazines and pays £6.
How many magazines did Kerry buy?

PS **7** A girl is Y years old. Her father is 23 years older than she is. The sum of their ages is 37. How old is the girl?

PS **8** A boy is X years old. His sister is twice as old as he is. The sum of their ages is 24. How old is the boy?

9 The diagram shows a rectangle.
Find x if the perimeter is 24 cm.

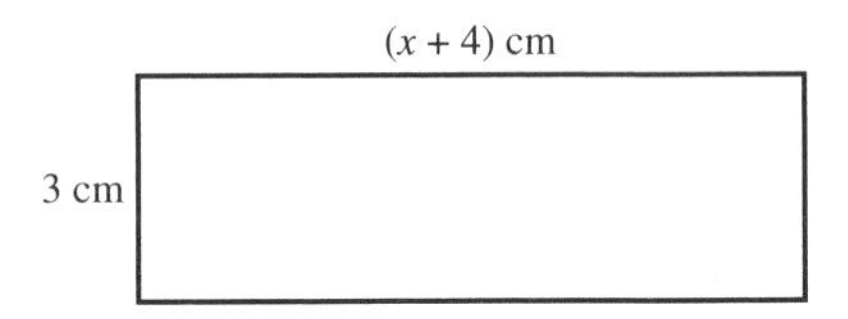

10 Find the length of each side of the pentagon, if it has a perimeter of 32 cm.

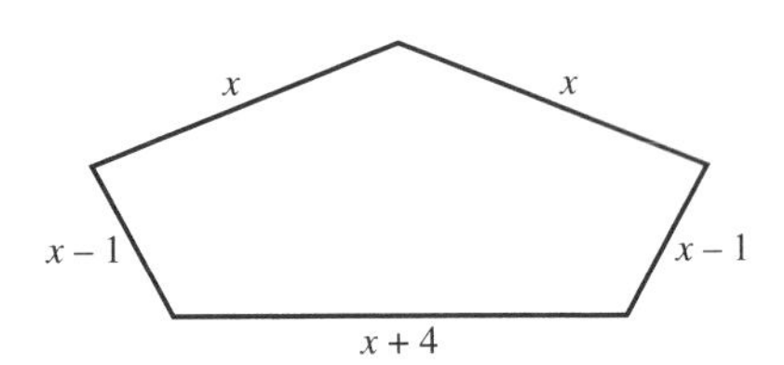

11 On a bookshelf there are $2b$ crime novels, $3b - 2$ science fiction novels and $b + 7$ romance novels. Find how many of each type of book there is, if there are 65 books altogether.

12 Maureen thought of a number. She multiplied it by 4 and then added 6 to get an answer of 26. What number did she start with?

13 Declan also thought of a number. He took away 4 from the number and then multiplied by 3 to get an answer of 24. What number did he start with?

PS **14** Sandeep's money box contains 50p coins, £1 coins and £2 coins.
In the box there are twice as many £1 coins than 50p coins and 4 more £2 coins than 50p coins. There are 44 coins in the box.

a Find how many of each coin there is in the box.

b How much money does Sandeep have in his money box?

FM **15** Olivia has some unlabelled tins of rice pudding.
She needs to find out how much they weigh.
Olivia doesn't have any weights but she does have a set of scales and some other tins with labels on them.
After some trial and improvement, she finds that five of her tins of rice and one tin of beans weighing 120 g balance with three of her rice tins and two jars of jam weighing 454 g each.
How much does one tin of rice weigh?

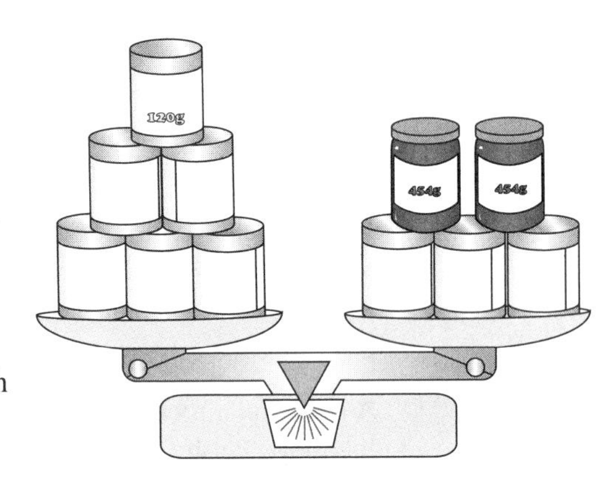

C

AU 16 The diagram shows two number machines that perform the same operations.

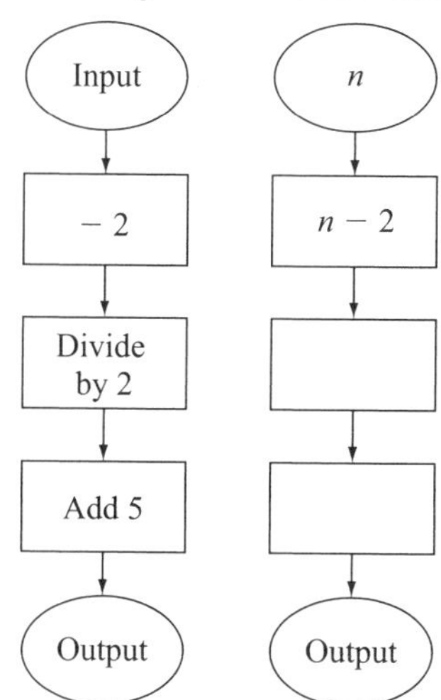

a Starting with an input value of 4, work through the left-hand machine to get the output.

b Find an input value that gives the same value for the output.

c Fill in the algebraic expressions in the right hand machine for an input of n (the first operation has been filled in for you).

d Set up an equation for the same input and output and show each step in solving the equation to get the answer in part **b**.

PS 17 Could the triangle shown here be an equilateral triangle?

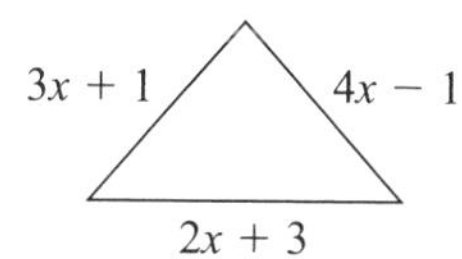

12.3 Trial and improvement

HOMEWORK 12F

C

1 Without using a calculator, find the two consecutive whole numbers between which the solution to each of the following equations lies.

a $x^3 = 10$ **b** $x^3 = 50$ **c** $x^3 = 800$
d $x^3 = 300$

2 Show that $x^2 + 2x = 20$ has a solution between $x = 3$ and $x = 4$, and find the solution to one decimal place.

3 Find two consecutive whole numbers between which the solution to each of the following equations lies.

a $x^3 + x = 7$ **b** $x^3 + x = 55$ **c** $x^3 + x = 102$
d $x^3 + x = 89$

4 Find a solution to each of the following equations to one decimal place.

a $x^3 - x = 30$ **b** $x^3 - x = 95$ **c** $x^3 - x = 150$
d $x^3 - x = 333$

5 Show that $x^3 + x = 45$ has a solution between $x = 3$ and $x = 4$, and find the solution to one decimal place.

C

6 Show that $x^3 - 2x = 95$ has a solution between $x = 4$ and $x = 5$, and find the solution to one decimal place.

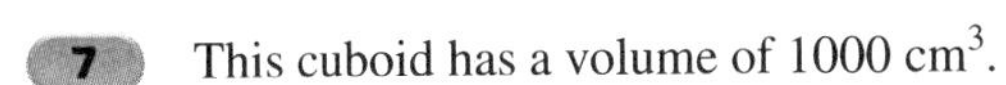

7 This cuboid has a volume of 1000 cm^3.

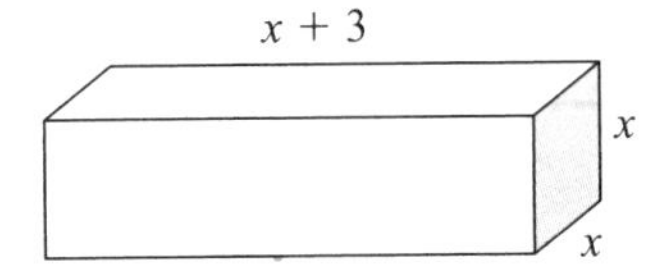

a Write down an expression for the volume.

b Use trial and improvement to find the value of x to one decimal place.

8 Darius is using trial and improvement to find a solution to the equation $x^3 - x^2 = 25$
This table shows his first trial.

x	$x^3 - x^2 = 25$	Comment
3	18	Too low

Copy and continue the table to find a solution to the equation.
Give your answer correct to one decimal place.

PS 9 Two numbers, a and b, are such that $ab = 20$ and $a - b = 5$
Use trial and improvement to find the two numbers to one decimal place.
Copy and then fill in the table below. The first two lines have been done for you.

a	$b = (20 \div a)$	$a - b$	Comment
5	4	1	Too low
10	2	8	Too high

12.4 Solving linear inequalities

HOMEWORK 12G

C

1 Solve the following linear inequalities.

a $x + 3 < 8$ **b** $t - 2 > 6$ **c** $p + 3 \geqslant 11$

d $4x - 5 < 7$ **e** $3y + 4 \leqslant 22$ **f** $2t - 5 > 13$

g $\dfrac{x+3}{2} < 8$ **h** $\dfrac{y+4}{3} \leqslant 5$ **i** $\dfrac{t-2}{5} \geqslant 7$

j $2(x - 3) < 14$ **k** $3x + 8 \geqslant 8$ **l** $4t - 1 \geqslant 29$

2 Write down the values of x that satisfy each of the following.

a $x - 2 \leqslant 3$, where x is a positive integer.

b $x + 3 < 5$, where x is a positive, odd integer.

c $2x - 14 < 38$, where x is a square number.

d $4x - 6 \leqslant 15$, where x is a positive, odd number.

e $2x + 3 < 25$, where x is a positive, prime number.

FM 3 Frank had £6. He bought three cans of cola and lent his brother £3. When he got home, he put a 50p coin in his piggy bank. What was the most that the cans of cola could have cost?

AU 4 The perimeter of this rectangle is bigger than 10 but less than 16.
What are the limits of the area?

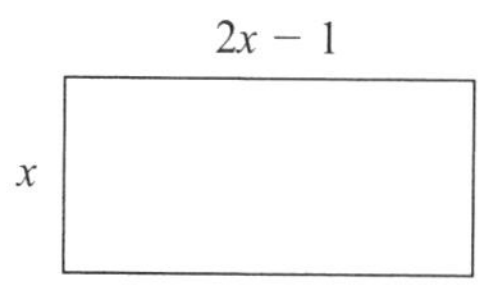

PS 5 A teacher asks six students to come to the front and hold up the following cards.

$x > 0$	$x < 2$	$x \geqslant 3$	$x = 2$	$x = 3$	$x < 9$

She writes 'TRUE' on one side of the board and 'FALSE' on the other side.
She asks other students to call out a number, and the students with the cards have to stand by the 'TRUE' side if their card is true for the number, or by the 'FALSE' side if it isn't.

a A student calls out '2' and the students all go to the correct side.
  i Which cards are held by the students on the 'TRUE' side?
  ii Which cards are held by the students on the 'FALSE' side?

b Find a value that would satisfy these groupings:

True

$x \geqslant 3$	$x < 9$	$x > 0$

False

$x < 2$	$x = 2$	$x = 3$

HOMEWORK 12H

1 Write down the inequality that is represented by each diagram below.

a

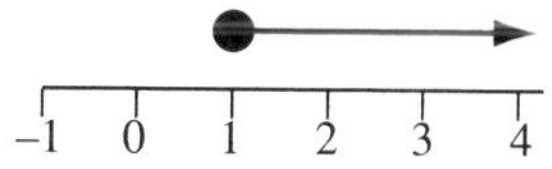

b

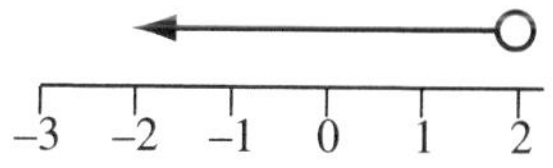

c

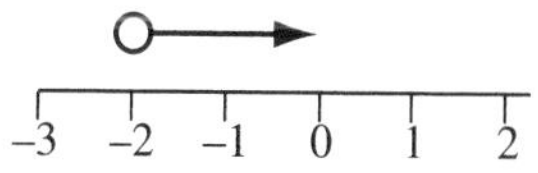

d

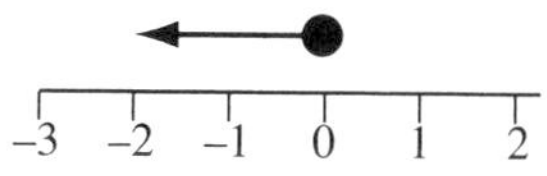

e

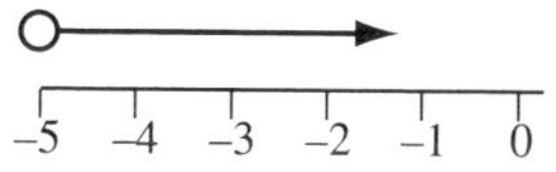

f

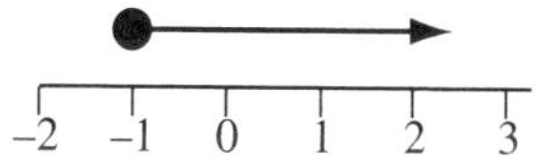

2 Draw diagrams to illustrate the following.

a $x \leqslant 2$ **b** $x > -3$ **c** $x \geqslant 1$ **d** $x < 4$

e $x \geqslant -3$ **f** $1 < x \leqslant 4$ **g** $-2 \leqslant x \leqslant 4$ **h** $-2 < x < 3$

3 Solve the following inequalities and illustrate their solutions on number lines.

a $x + 5 \geqslant 9$ **b** $x + 4 < 2$ **c** $x - 2 \leqslant 3$ **d** $x - 5 > -2$

e $4x + 3 \leqslant 9$ **f** $5x - 4 \geqslant 16$ **g** $2x - 1 > 13$ **h** $3x + 6 < 3$

i $2x + 1 < 15$ **j** $\frac{x+1}{2} \leqslant 2$ **k** $\frac{x-3}{3} > 7$ **l** $\frac{x+6}{4} \geqslant 1$

FM 4 Mary went to the record shop with £20. She bought two CDs costing £x each and a DVD costing £9.50. When she got to the till, she found she didn't have enough money.
Mary left the DVD and paid for the two CDs.
On the way home, she had enough money to buy a lipstick for £7.

a Explain why $2x + 9.5 > 20$ and solve the inequality.
b Explain why $2x + 7 \leqslant 20$ and solve the inequality.
c Show the solution to both of these inequalities on a number line.
d What is the price of a CD if it is a whole number of pounds?

C

AU 5 Copy the number lines below and draw two inequalities on them so that they have the integers {5, 6, 7, 8} in common.

0 1 2 3 4 5 6 7 8 9 10 x

0 1 2 3 4 5 6 7 8 9 10 x

12.5 Drawing quadratic graphs

HOMEWORK 12I 

C

1 **a** Copy and complete the table for the graph of $y = 2x^2$ for $-3 \leqslant x \leqslant 3$.

x	–3	–2	–1	0	1	2	3
$y = 2x^2$	18		2			8	

b Use the graph to find the value of y when $x = -1.4$.

c Use the graph to find the values of x that give a y-value of 10.

2 **a** Copy and complete the table for the graph of $y = x^2 + 3$ for $-5 \leqslant x \leqslant 5$.

x	–5	–4	–3	–2	–1	0	1	2	3	4	5
$y = x^2 + 3$	28		12					7			28

b Use the graph to find the value of y when $x = 2.5$.

c Use the graph to find the values of x that give a y-value of 10.

3 **a** Copy and complete the table for the graph of $y = x^2 - 3x + 2$ for $-3 \leqslant x \leqslant 4$.

x	–3	–2	–1	0	1	2	3	4
$y = x^2 - 3x + 2$	20			2			2	

b Use the graph to find the value of y when $x = -1.5$.

c Use the graph to find the values of x that give a y-value of 2.5.

4 **a** Copy and complete the table to draw the graph of $y = x^2 - 5x + 4$ for $-1 \leqslant x \leqslant 6$.

x	–1	0	1	2	3	4	5	6
$y = x^2 - 5x + 4$	10	4				0		

b Where does the graph cross the x-axis?

c Use your graph to find the y-value when $x = 2.5$.

d Use your graph to find the values of x that give a y-value of 8.

PS 5 Tom is drawing quadratic equations of the form $y = x^2 + bx + c$
He notices that two of his graphs pass through the point (2, 5).
Which of the following equations are of the two graphs?

Equation A: $y = x^2 + 3$
Equation B: $y = x^2 + 1$
Equation C: $y = x^2 + 2x - 3$
Equation D: $y = x^2 - x + 5$

AU 6 All parabolas have a special property.
If a light is placed at a point called the 'focus', marked F on the y-axis, any light from the focus hitting the parabola is reflected parallel to the y-axis.

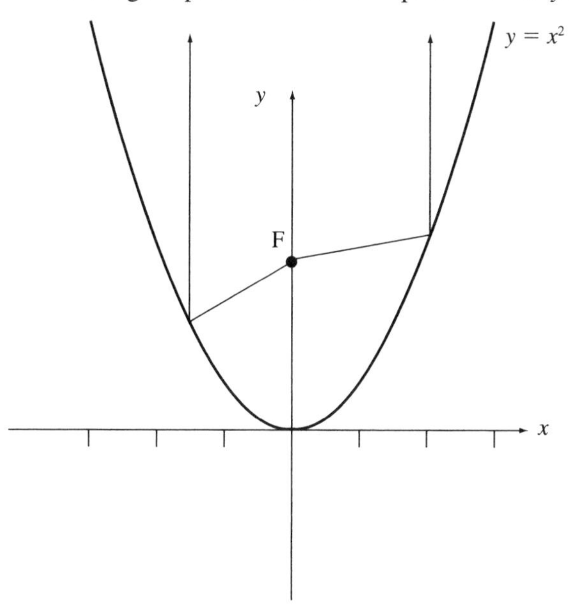

Explain how this might have a practical use in cars.

Problem-solving Activity

Drawing quadratic graphs

The sketch below shows the graphs of $y = x^2$ and $y = x^2 + 3x$.

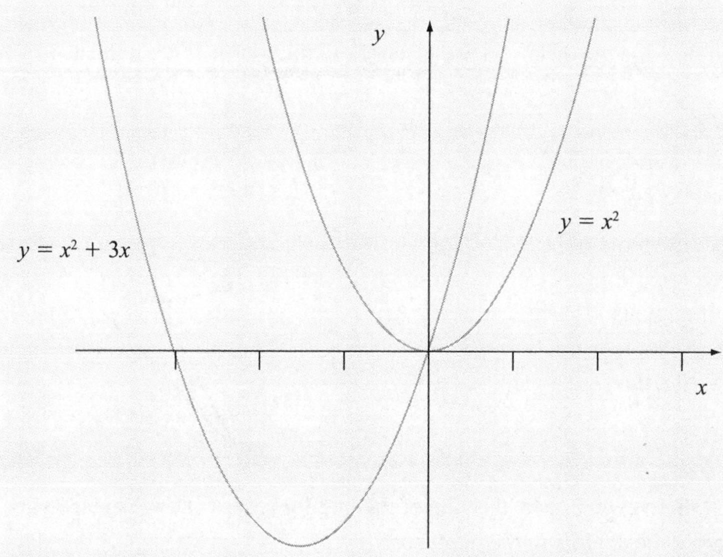

Use a graphical calculator or a computer graphing programme to draw the graphs of $y = x^2$ and $y = x^2 + 3x$ on the same axes.

Now draw graphs of $y = x^2 + 4x$, $y = x^2 - 2x$ and $y = x^2 - x$ on the same axes.

Describe anything that you notice about these graphs.

What effect does adding or subtracting a term in x have on the graph of $y = x^2$?

Where do these graphs cross the x-axis?

13 Geometry and measures: Angles and bearings

13.1 Angles in a polygon

HOMEWORK 13A 

E

1 Find the size of the angle marked with a letter in each of these quadrilaterals.

a

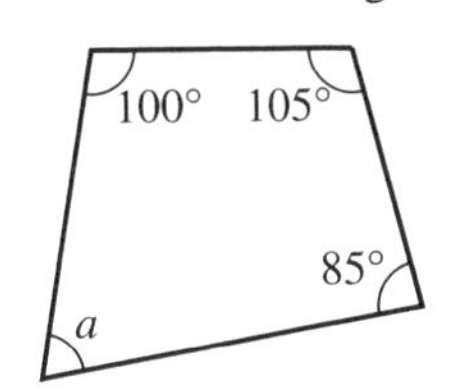

b

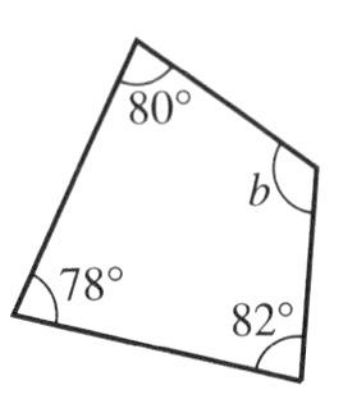

c

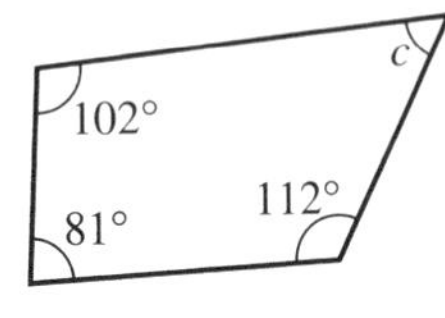

d

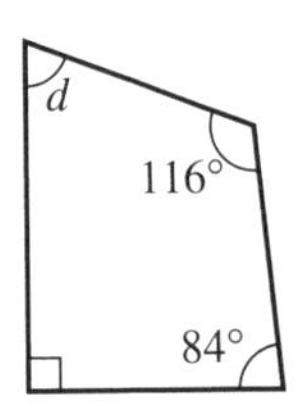

e

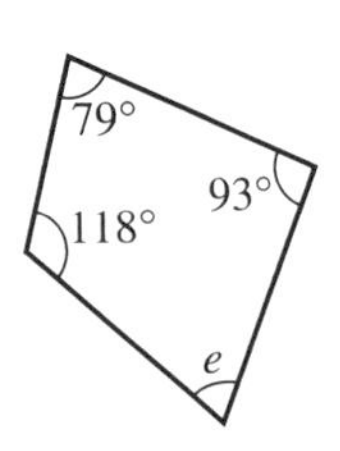

f

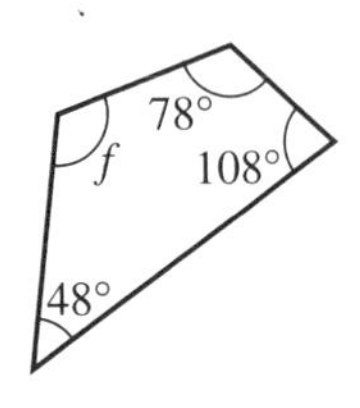

AU 2 State whether each of these sets of angles are the four interior angles of a quadrilateral? Explain your answers.

a 125°, 65°, 70° and 90°
b 100°, 60°, 70° and 130°
c 85°, 95°, 85° and 95°
d 120°, 120°, 70° and 60°
e 112°, 68°, 32° and 138°
f 151°, 102°, 73° and 34°

3 Three interior angles of a quadrilateral are given. Find the fourth angle indicated by a letter.

a 110°, 90°, 70° and a°
b 100°, 100°, 80° and b°
c 60°, 60°, 160° and c°
d 135°, 122°, 57° and d°
e 125°, 142°, 63° and e°
f 102°, 72°, 49° and f°

4 For the quadrilateral on the right:

a find the size of angle x.
b What type of angle is x?
c What is the special name of a quadrilateral like this?

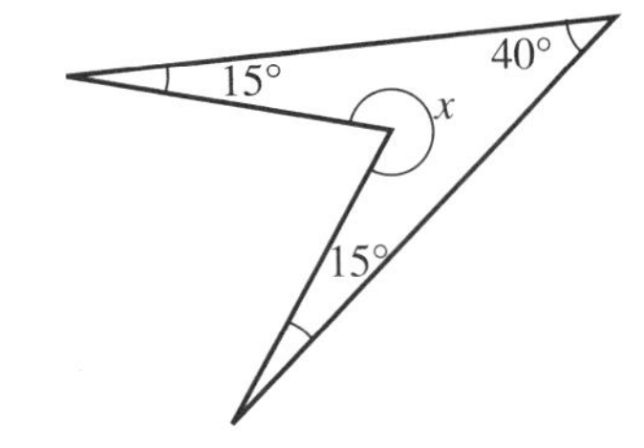

D

AU 5 **a** Draw a diagram to explain why the sum of the interior angles of any pentagon is 540°.
b Find the size of the angle x in the pentagon.

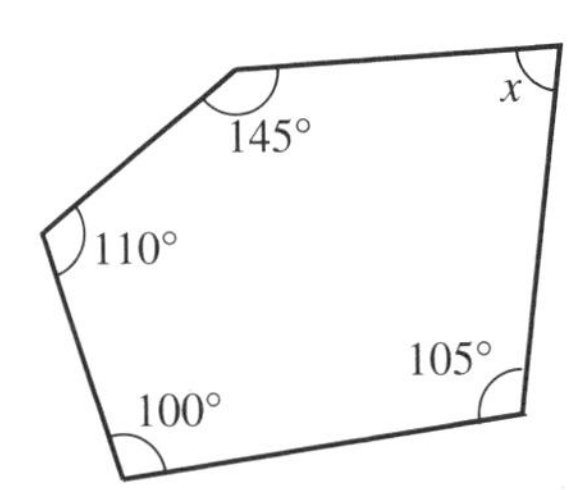

D

6 Calculate the size of the angle marked with a letter in each of the polygons below.

a

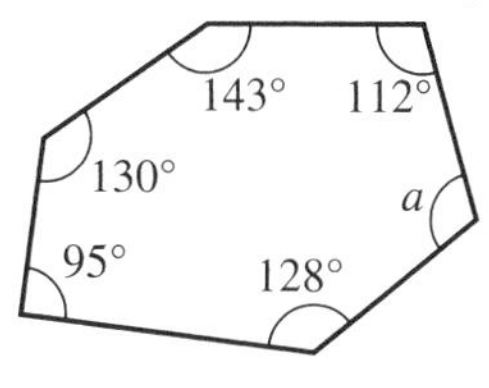

b

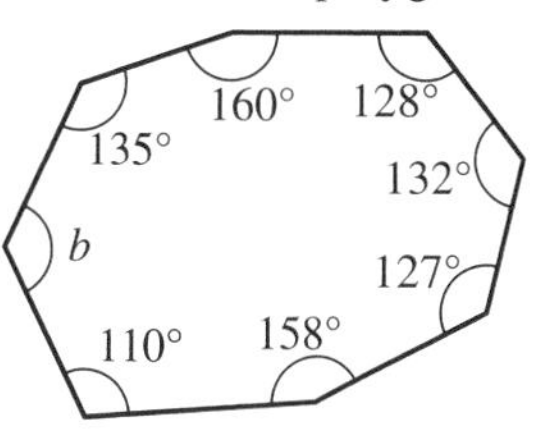

FM 7 Jamal is cutting metal from a rectangular sheet to make this sign.

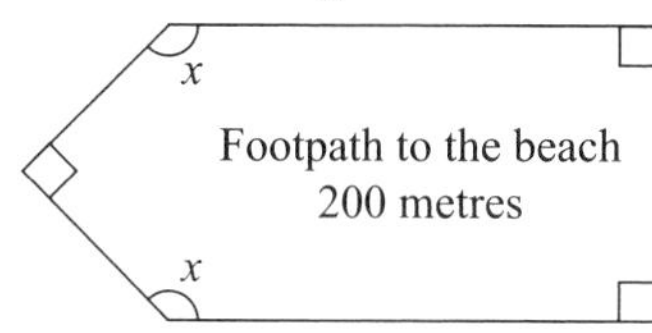

He needs to cut the two angles marked x accurately. How big is each one?

PS 8 Find the value of x in the diagram.

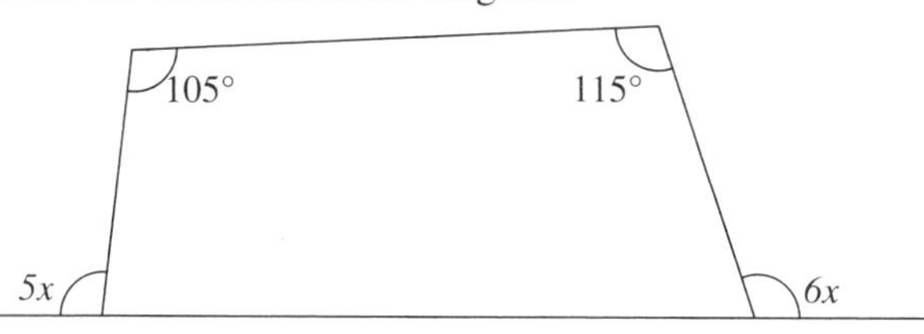

AU 9 The diagram shows a quadrilateral.

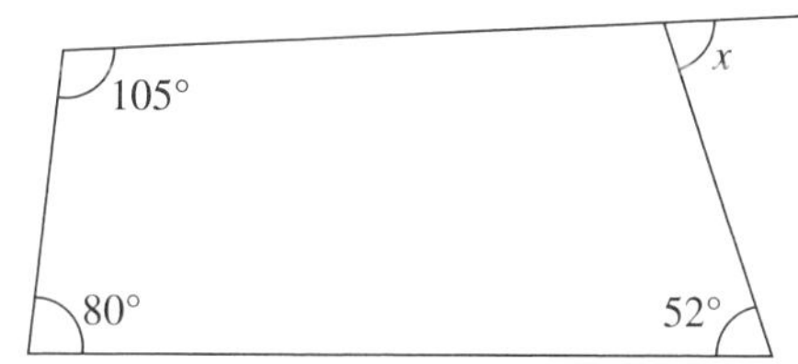

Paul says that the size of angle x is 52°
Explain why Paul is wrong.
Work out the correct value for x.

13.2 Regular polygons

HOMEWORK 13B

C

1 For each regular polygon below, find the interior angle x and the exterior angle y.

a

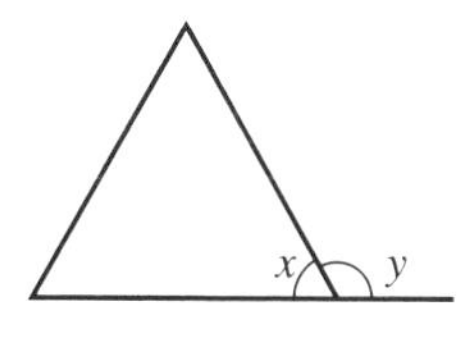

b

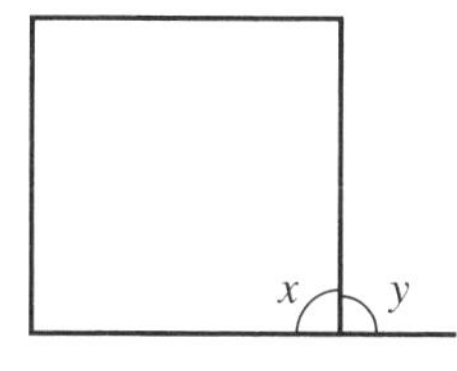

c

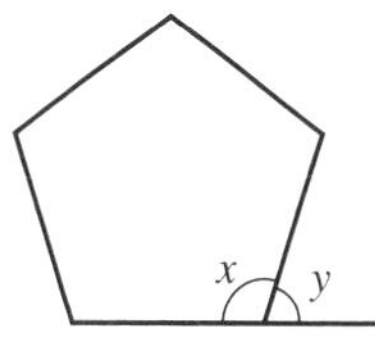

d

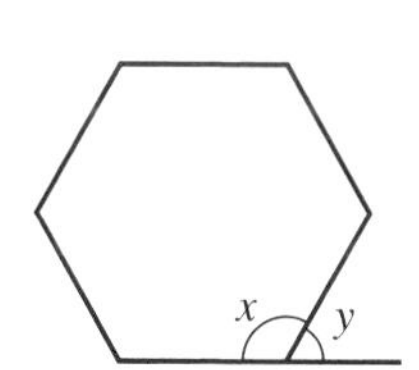

e

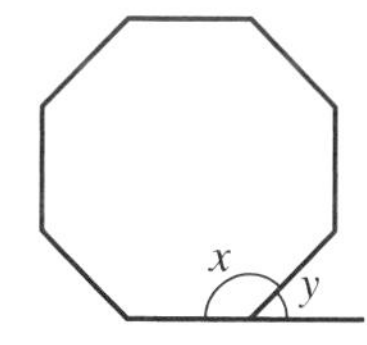

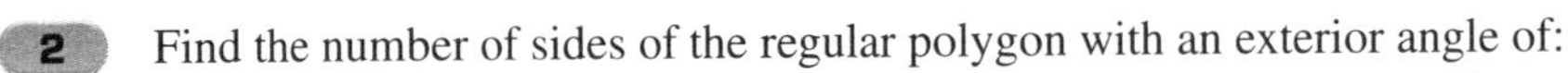

C

2 Find the number of sides of the regular polygon with an exterior angle of:

a 20° **b** 30° **c** 18° **d** 4°.

3 Find the number of sides of the regular polygon with an interior angle of:

a 135° **b** 165° **c** 170° **d** 156°.

4 What is the name of the regular polygon whose interior angle is treble its exterior angle?

FM 5 Four regular octagonal tiles of the same size are put together to make a floor tiling pattern.

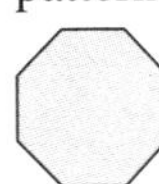 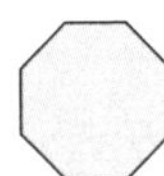 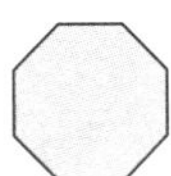 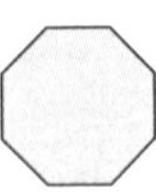

What is the shape of the tile that is required to fill the gap?

PS 6 ABCDE is a regular pentagon.

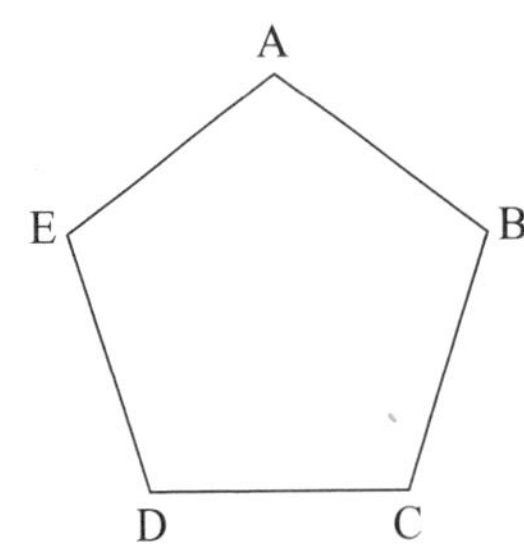

Work out the size of angle ADE.
Give reasons for your answer.

AU 7 Which of the following statements are true for a regular hexagon?

1. The size of each interior angle is 60°
2. The size of each interior angle is 120°
3. The size of each exterior angle is 60°
4. The size of each exterior angle is 240°

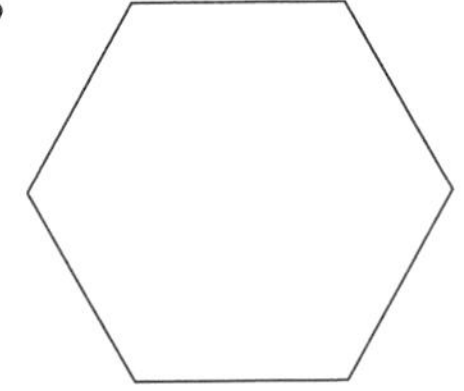

13.3 Bearings

HOMEWORK 13C

D

1 **a** Write down the bearing of B from A. **b** Write down the bearing of D from C.

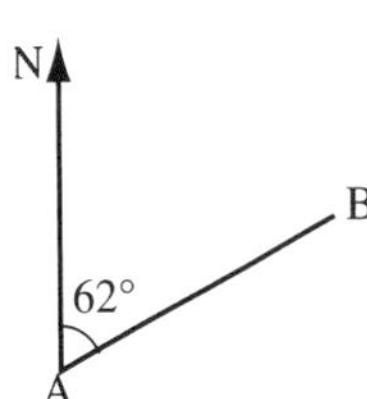

c Write down the bearing of F from E. **d** Write down the bearing of H from G.

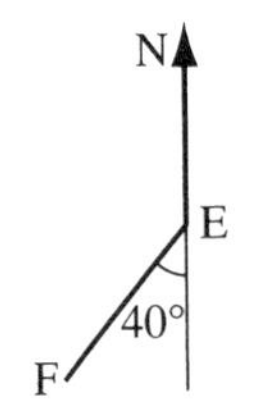

N
H
75°
G

D

2 On the right is a map of Britain.
By measuring angles, find the bearings of:

a London from Edinburgh
b London from Cardiff
c Edinburgh from Cardiff
d Cardiff from London.

FM 3 An aircraft flies directly from London to Paris.
The diagram shows the aircraft's flight path.

a Write down the three-figure bearing of Paris from London.
b Work out the actual distance from London to Paris.
c The aircraft flies directly back to London.
What is the three-figure bearing of London from Paris?

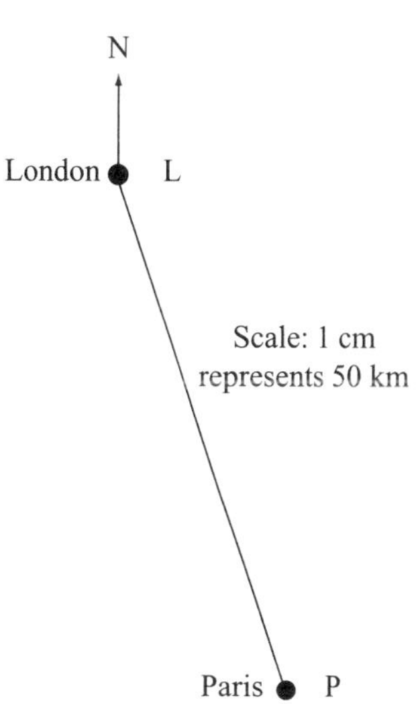

PS 4 **a** The bearing of B from A is $x°$.
What is the bearing of A from B?

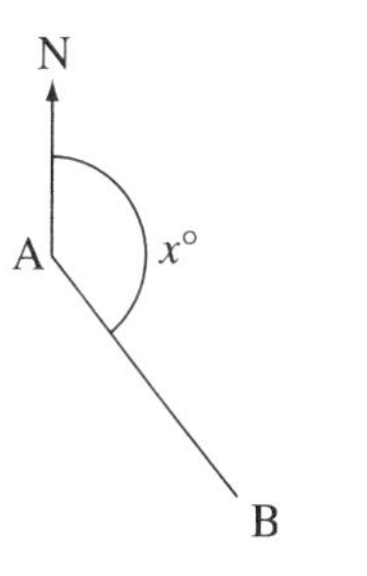

b The bearing of Y from X is $y°$.
What is the bearing of X from Y?

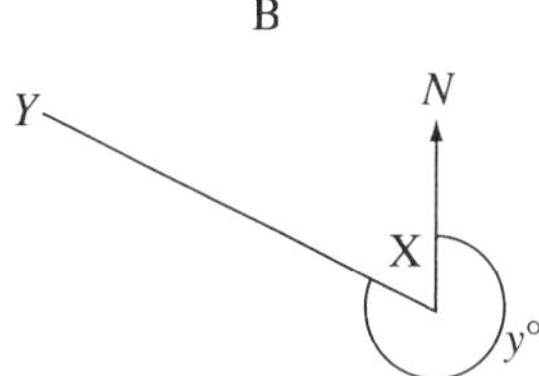

5 Town B is 40 km from Town A and on a bearing of 050°.
Town C is 60 km from Town A and on a bearing of 300°.
Make a scale drawing to find the bearing of Town B from Town C.

FM 6 A ship is sailing at a bearing of 324° when it receives an order to change course to due East. Through how many degrees should it turn?

PS 7 The three towns, A, B and C, form an equilateral triangle. B is due north of A. The bearing of C from A is 060°. What is the bearing of C from B?

Functional Maths Activity

Back bearings

The three-figure bearing of B from A is 060°.

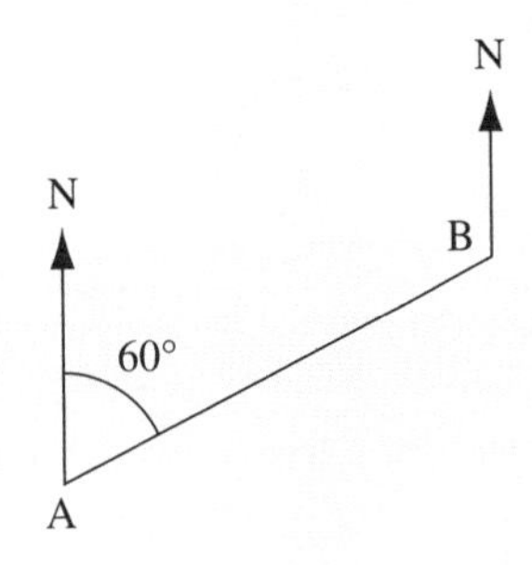

The three-figure bearing of A from B is known as a *back bearing*.

The diagram below shows that the back bearing of A from B is 240°.

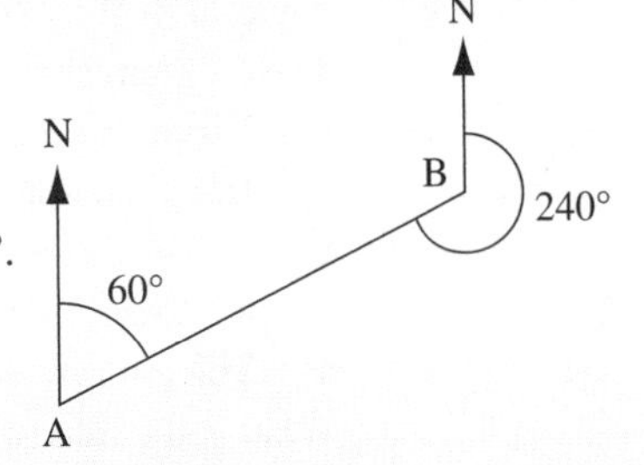

Trevor is taking an examination for his pilot's licence. Here is one of his questions.

The map shows three airports in England.

a Use the map to find the three-figure bearing for each of the following flights:

i Newcastle from Manchester

ii Heathrow from Newcastle

iii Manchester from Heathrow.

b Calculate the back bearing for each flight.

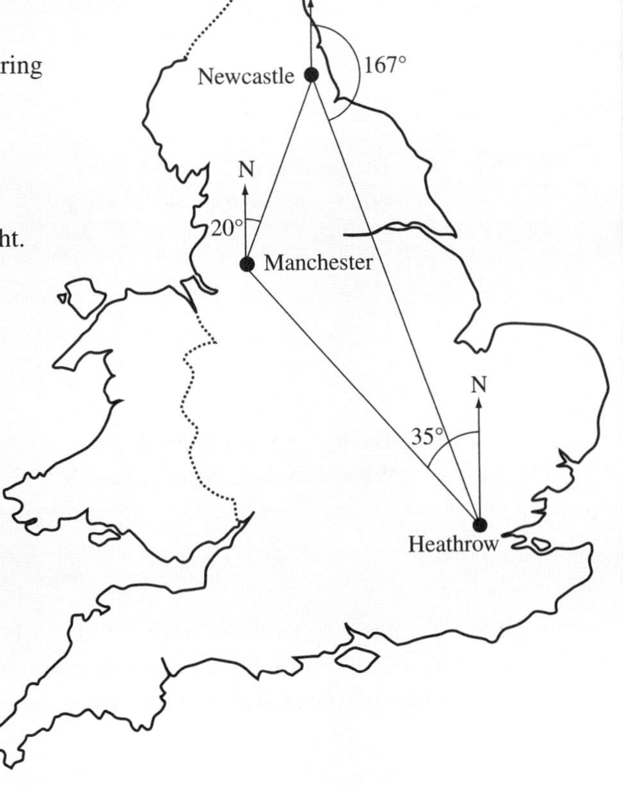

14 Geometry: Circles

14.1 Drawing circles

HOMEWORK 14A

1 Measure the radius of each of the following circles, giving your answers in centimetres. Write down the diameter of each circle.

a

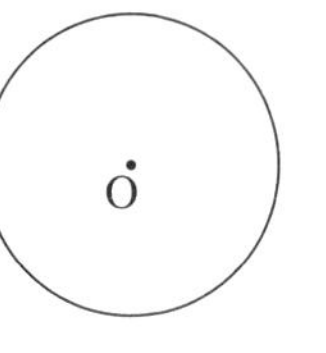

b

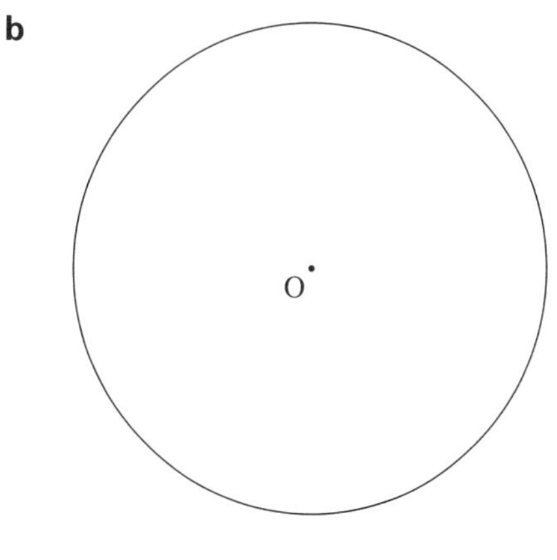

c

O

2 Draw circles with the following measurements.

a Radius = 1.5 cm **b** Radius = 4 cm

c Diameter = 7 cm **d** Diameter = 9.6 cm

3 Accurately draw the following shapes.

a

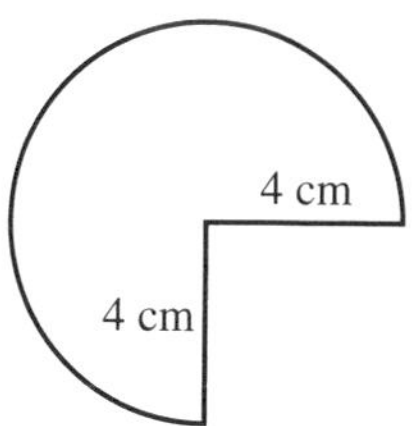

b

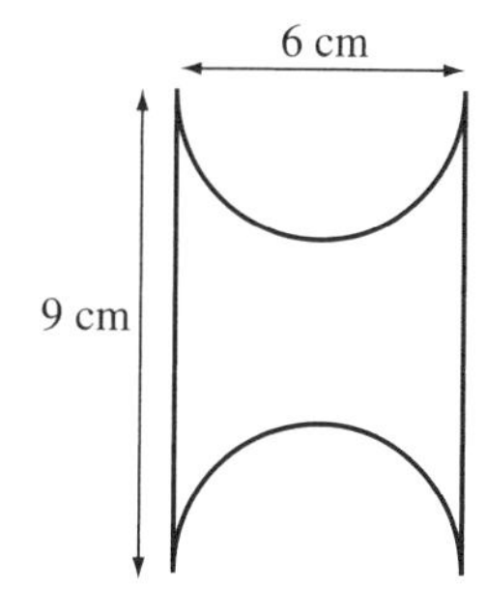

F

4 Draw an accurate copy of this diagram.
What is the length of the diameter of the circle?

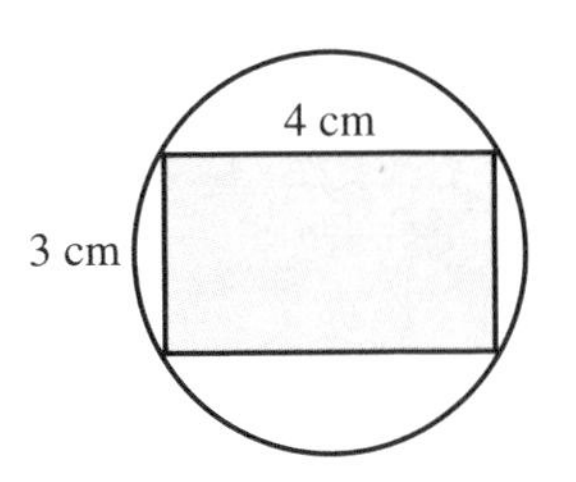

5 Draw an accurate copy of this diagram.
What is the diameter of the circle?

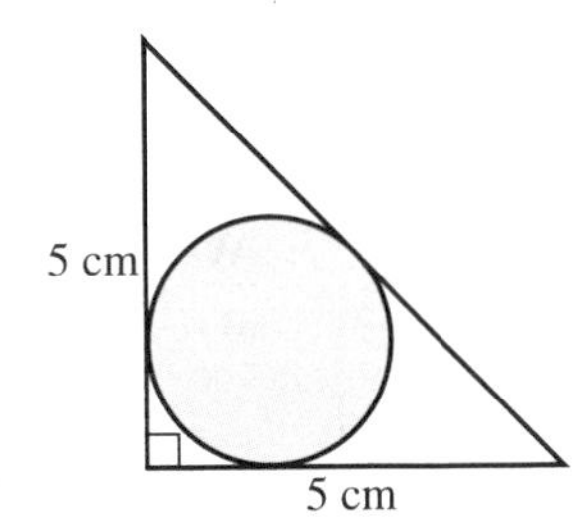

PS 6 Six identical circles fit inside a rectangle.

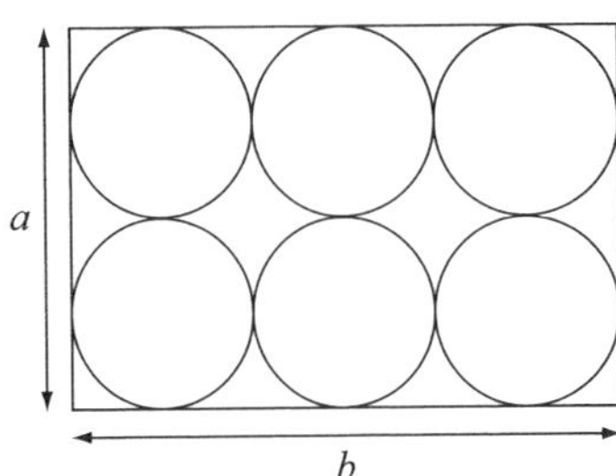

The radius of each circle is 3 cm.
Work out the lengths of *a* and *b* marked on the diagram.

AU 7 The diagram shows a circle with centre O.
The line XYZ touches the circle at Y.

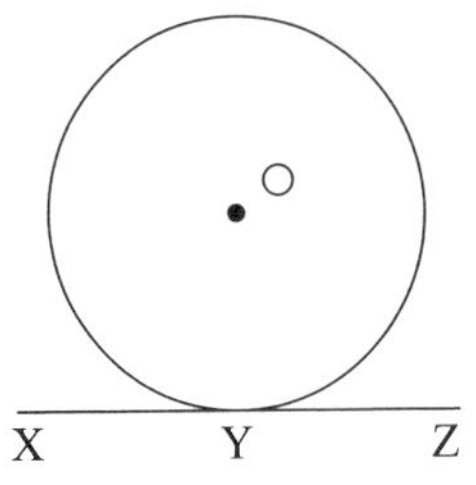

a Write down the mathematical name of the line XYZ.
b On a copy of the diagram, draw the radius that meets the circumference at Y.
c What do you notice about the angle OYZ?

FM 8 Coins with a radius of 1 cm are cut from a strip of metal, which is 2 cm wide and 45 cm long. How many coins can be cut from the strip?

14.2 The circumference of a circle

HOMEWORK 14B

Example Calculate the circumference of the circle with a diameter of 4 cm.

Use the formula $c = \pi d$. So $c = \pi \times 4 = 12.6$ cm (1dp).

1 Calculate the circumference of each circle illustrated below.
Give your answers to one decimal place.

a
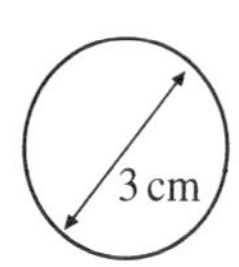

b
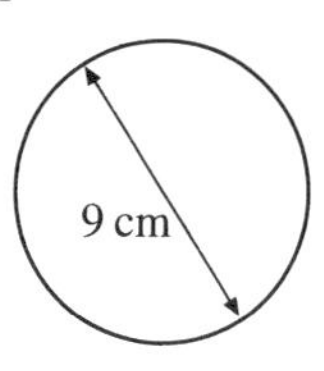

c
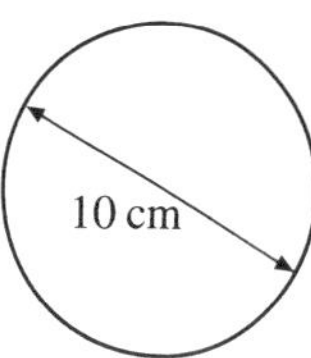

d
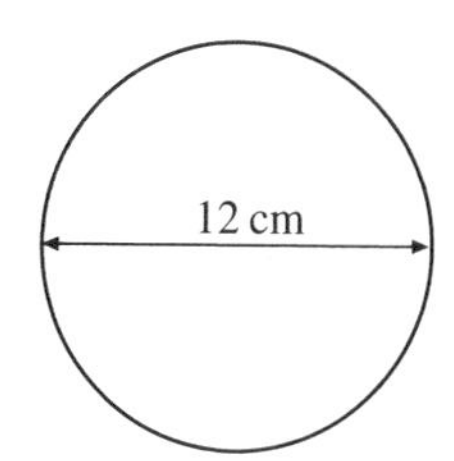

e
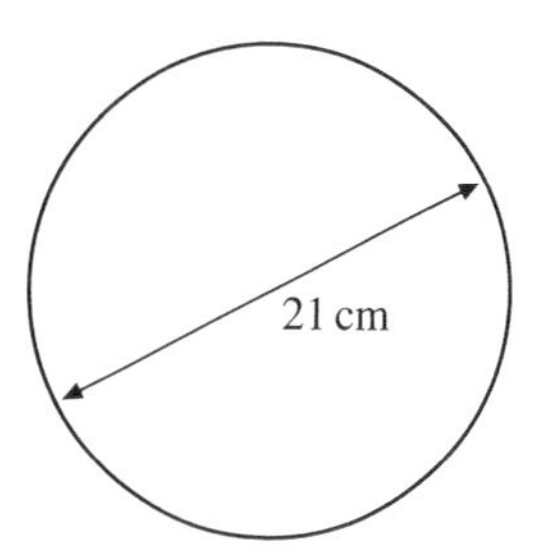

2 Calculate the circumference of each circle illustrated below.
Give your answers to one decimal place.

a
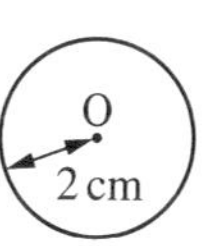

b
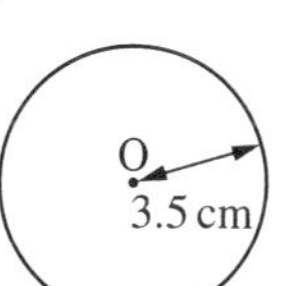

c
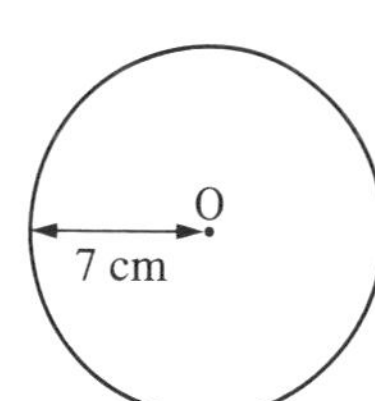

d
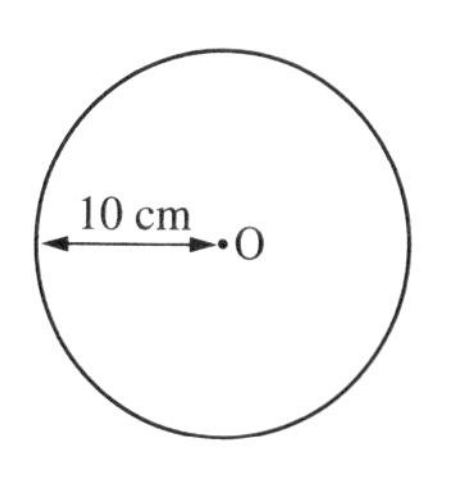

e
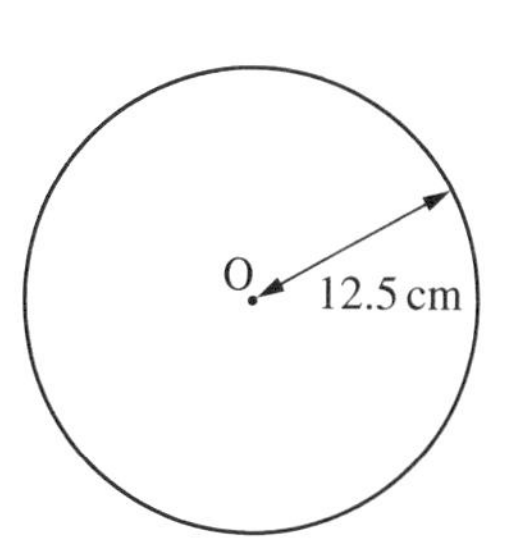

FM 3 Pat needs to put a fence around her circular pond, which has a diameter of 15 metres. What is the length of fencing she requires, if she buys the fencing in one-metre lengths?

FM 4 Roger trains for an athletics competition by running around a circular track, which has a radius of 50 m.

a Calculate the circumference of the track. Give your answer to one decimal place.

b How many complete circuits will he need to run to be sure of running 5000 m?

D

C

5 Calculate the perimeter of this semicircle.

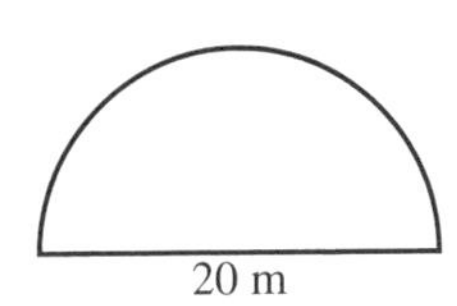

6 What is the diameter of a circle with a circumference of 40 cm? Give your answer to one decimal place.

AU 7 A trundle wheel is used by surveyors to estimate distances. A click is made every time the wheel travels one metre along the ground.

If you wanted to make a trundle wheel of your own, what would the radius have to be?

PS 8 A circle has a radius of r cm. Another circle has a radius of $(r + 1)$ cm.

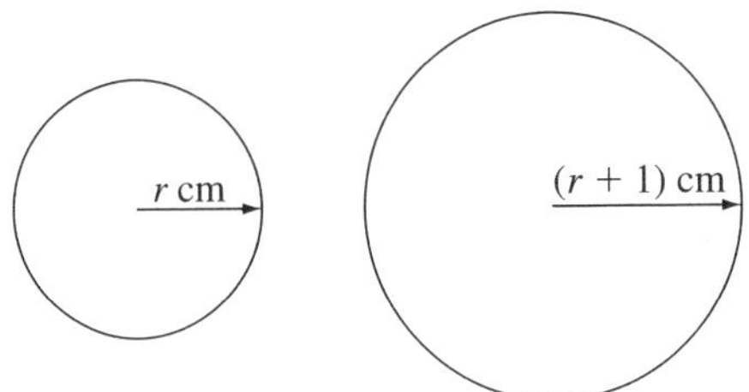

Prove that the difference in the two circumferences is 2π cm.

FM 9 The diameter of a cotton reel is 3 cm.

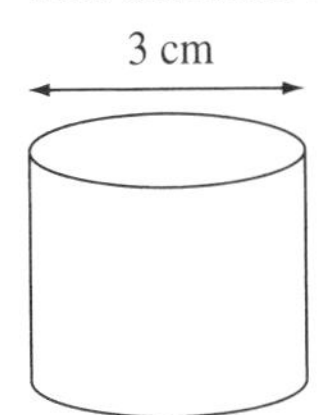

Cotton is wound onto the reel by rotating it on a machine.
A manufacturer wants to put on 80 m of cotton. How many rotations should he set the machine for?

14.3 The area of a circle

HOMEWORK 14C 

Example Calculate the area of a circle with a radius of 7 cm.

Use the formula $A = \pi r^2$.

So $A = \pi \times r \times r = \pi \times 7 \times 7 = 153.9\ \text{cm}^2$ (1dp).

1 Calculate the area of each circle illustrated below. Give your answers to one decimal place.

a
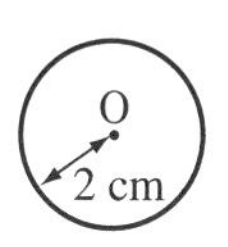

b
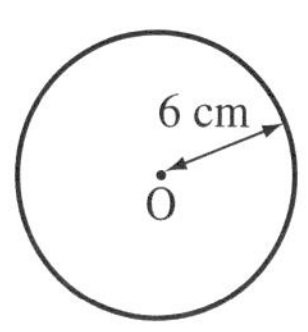

c
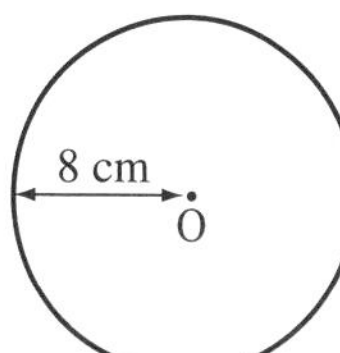

d
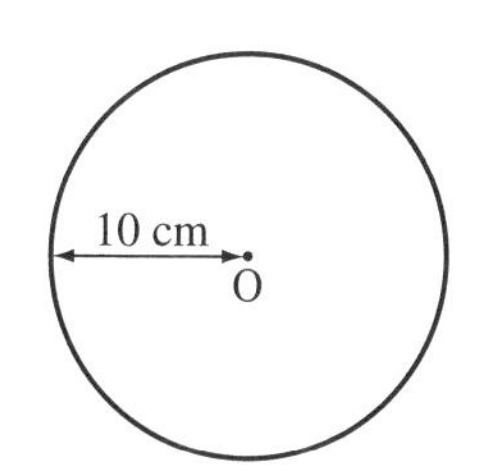

e
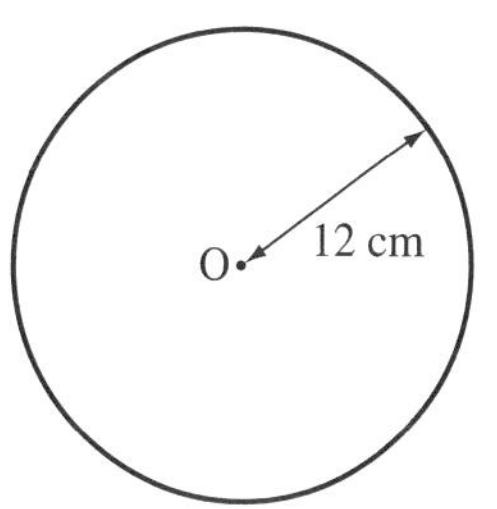

2 Calculate the area of each circle illustrated below. Give your answers to one decimal place.

a
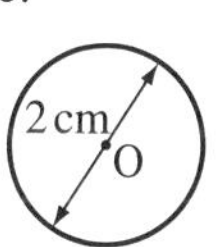

b
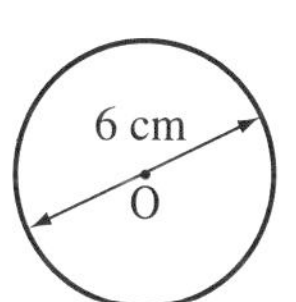

c
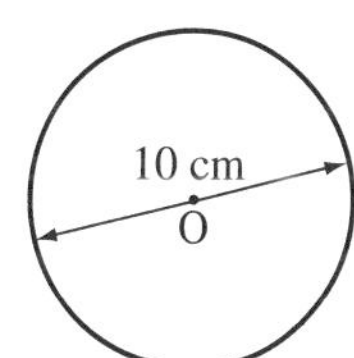

d
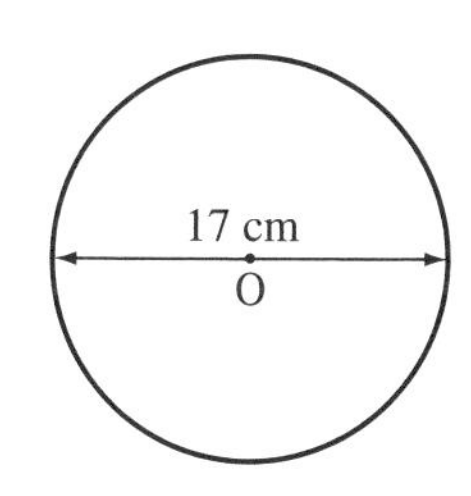

e
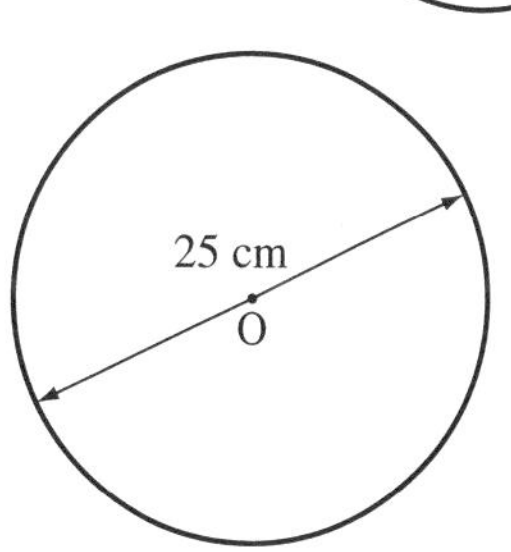

FM 3 A circular table has a diameter of 80 cm.

a Helen is preparing a meal for six people. She knows that to sit in comfort around the table, each person needs 30 cm of space. Is the table big enough for six people? Give a reason for your answer.

b A tablecloth should have an overlap of about 10 cm. What size of circular tablecloth is needed for this table?

C

4 The diagram shows a circular path around a flower bed in a park. The radius of the flower bed is 6 m and the width of the path is 1 m.

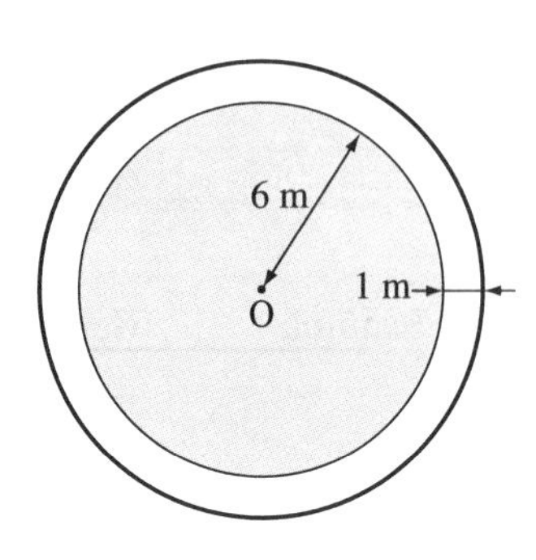

a Calculate the area of the flower bed.

b Write down the radius of the large circle.

c Calculate the area of the large circle.

d Calculate the area of the path.

FM **e** A company charges £12 per square metre for concrete. Dan wants a new path and has £300 to spend. Can he afford a concrete path?

5 The diagram shows a running track.

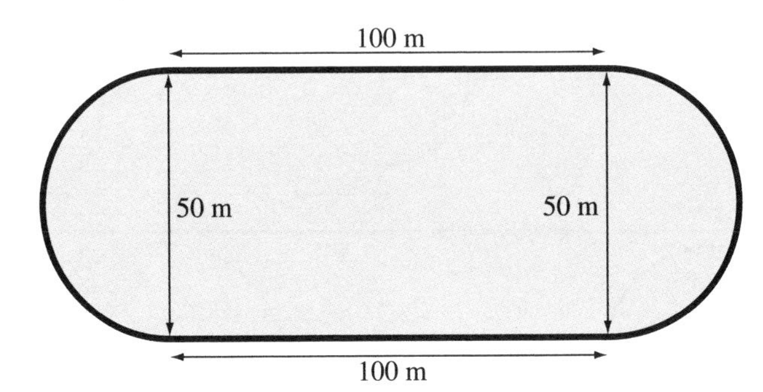

a Calculate the perimeter of the track. Give your answer to the nearest whole number.

b Calculate the total area inside the track. Give your answer to the nearest whole number.

6 A circle has a circumference of 50 cm.

a Calculate the diameter of the circle to one decimal place.

b What is the radius of the circle to one decimal place?

c Calculate the area of the circle to one decimal place.

7 The diagram shows a metal ring used by a manufacturing company.

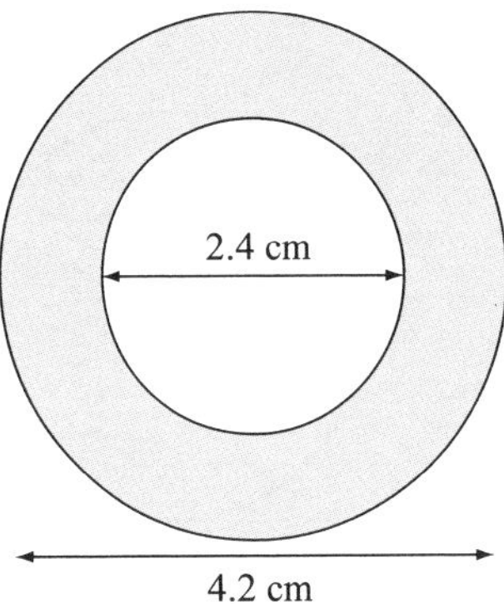

Calculate the area of the ring. Give your answer to one decimal place.

AU **8** The diameter of a circle is d.

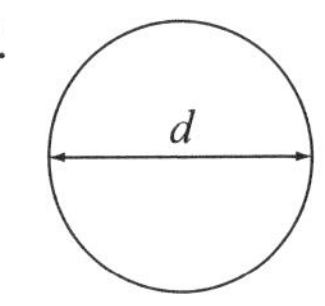

Engineers often use the formula $A = \frac{\pi d^2}{4}$ for the area of a circle instead of $A = \pi r^2$.

Can you show that these give the same answer?

C

PS 9 The diagram shows a shape made from a semicircle and a rectangle.

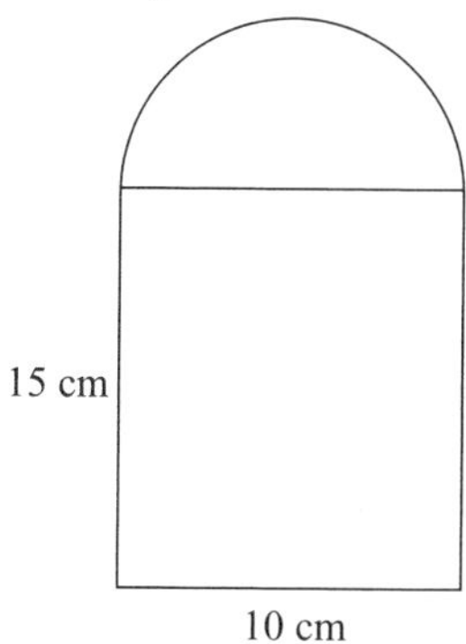

What is the area of this shape?

14.4 Answers in terms of π

HOMEWORK 14D

Example Write down the circumference (c) and area (A) of a circle whose radius is 5 cm. Give your answers in terms of π.

$c = \pi d = \pi \times 10 = 10\pi$ cm

$A = \pi r^2 = \pi \times 25 = 25\pi$ cm^2

Leave all your answers in terms of π.

D

1 State the circumference of the following circles.

a Diameter 7 cm **b** Radius 5 cm **c** Diameter 19 cm **d** Radius 3 cm

2 State the area of the following circles.

a Radius 8 cm **b** Diameter 7 cm **c** Diameter 18 cm **d** Radius 9 cm

FM **3** The diagram shows a circular pond with a diameter of 1.8 m.

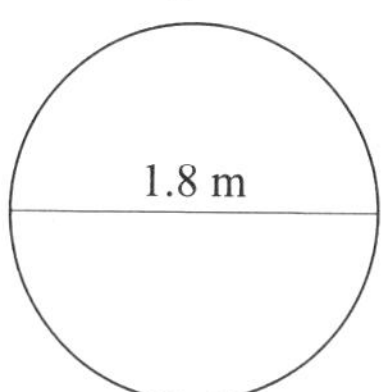

Sasha wants to put water lilies in the pond. This is done by putting pots under the water with individual plants in. She wants to have about six plants per square metre.
How many should she buy?

AU **4** Sean is working out the area of this circle, which has a radius of 8 cm.
He is going to write the area down in terms of π.
He writes down:
$A = 16\pi$ cm^2
Explain why Sean is wrong.

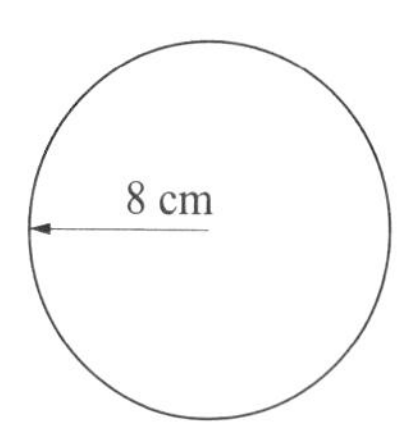

C

5 State the diameter of a circle with a circumference of 4π cm.

6 State the radius of a circle with an area of 36π cm^2.

7 State the diameter of a circle with a circumference of 20 cm.

8 State the radius of a circle with an area of 20 cm^2.

9 Calculate **i** the perimeter and **ii** the area for each of the following shapes, giving your answers in terms of π.

a

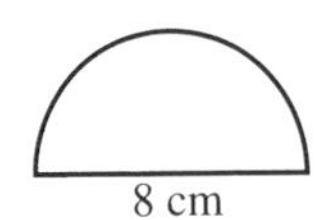

b

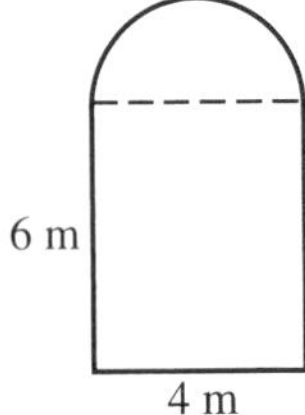

PS 10 A star shape is made by cutting four quadrants from a square with side length $2a$.

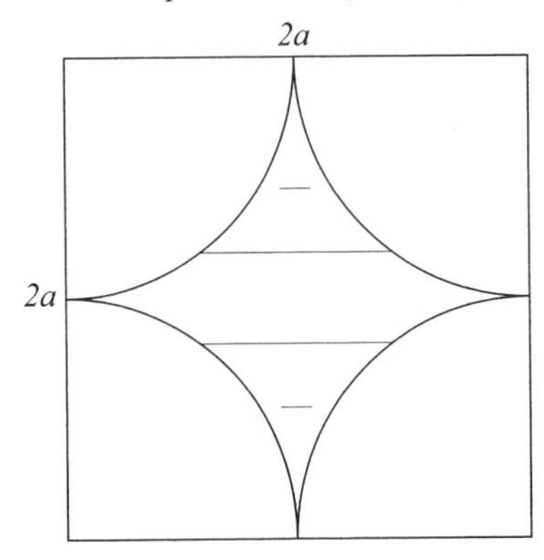

Find a formula for the area of the star in terms of a.

Functional Maths Activity

Track-and-field event measurement

The diagram shows the layout of a sports field for track-and-field events. Joe is a University student of Sport Science and has been set a task to answer the questions below about track-and-field events. Help him to answer the questions.

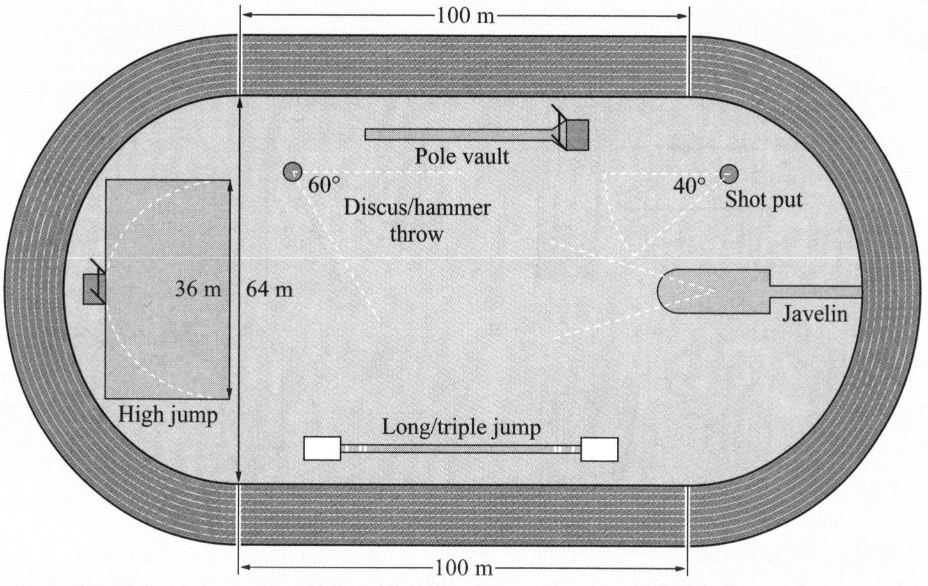

1 Calculate the area of the space inside the running track.
Give your answer to the nearest 10 m^2.

2 Calculate the area of the semicircle inside the high jump zone.
Give your answer to the nearest m^2.

3 Calculate the total distance around the inner marking of the running track.
Give your answer to the nearest metre.

4 The width of each lane is 1.25 m.
Calculate the total distance around the outer marking of the running track.
Give your answer to the nearest metre.

5 An athlete completes a 100 m sprint in 12 seconds. Calculate his speed in kilometres per hour.

6 Why is a track field designed in an oval shape, rather than in a circle?

15 Geometry: Transformations

15.1 Congruent shapes

HOMEWORK 15A

G

1 State whether each pair of shapes **a** to **f** are congruent or not.

a

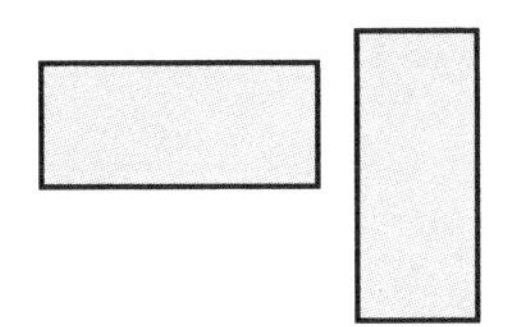

b

c

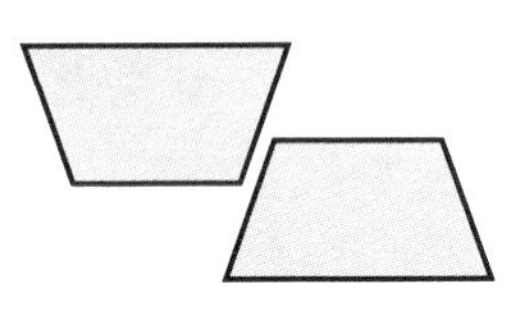

d

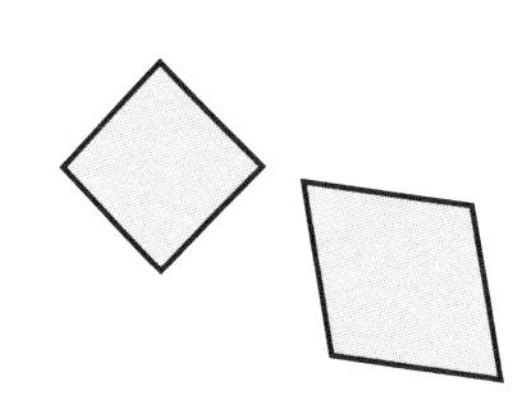

e

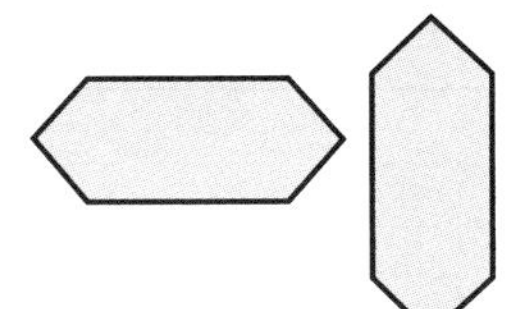

f

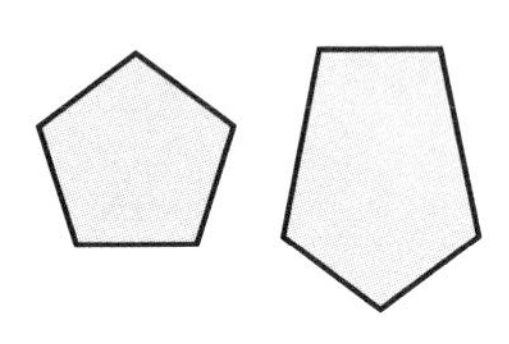

2 Which figure in each group of shapes is not congruent to the other two?

a

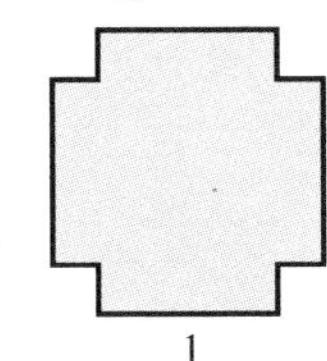

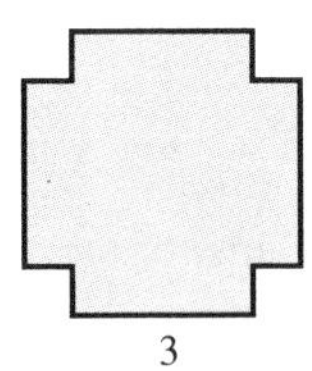

1 2 3

b

1 2 3

c

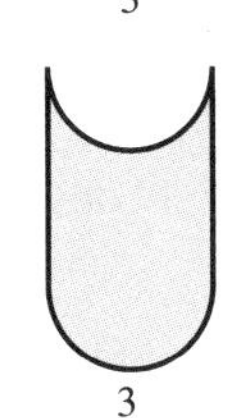

1 2 3

d

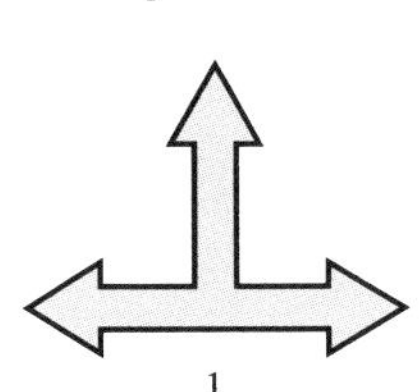

1 2 3

FM 3 While she was sitting in a doctors waiting room, Katrina looked at a simple design made of wood built into the wall as shown.

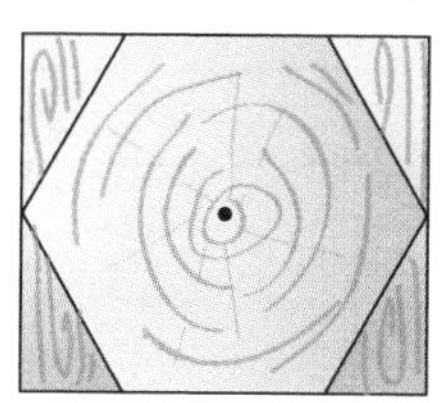

a What is the name of this six-sided regular shape?

b If you were to label its vertices ABCDEF and the centre O, how many other shapes inside the six-sided shape are congruent to:

i triangle ABC? **ii** quadrilateral ABCD?

iii triangle ACE? **iv** rectangle ABDE?

v triangle AOB? **vi** triangle ABD?

vii pentagon ABCDE?

4 The kite ABCD is shown on the right. The diagonals AC and BD intersect at X.

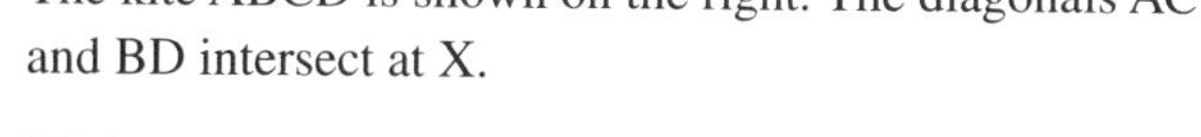

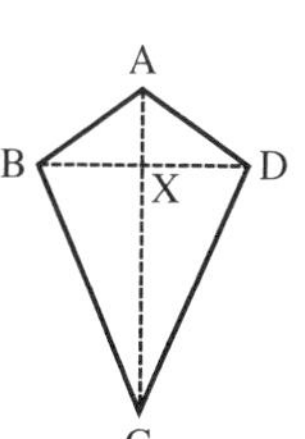

Which of the following statements are true?

a Triangle ABC is congruent to triangle ACD.

b Triangle ABD is congruent to triangle BCD.

c Triangle XBC is congruent to triangle XCD.

PS 5 There are two single-digit numbers that could be considered congruent.
How many numbers less than one hundred can you write down that could be considered congruent with another?
Write all these numbers down in their congruent pairs.

AU 6 Uzma and Pete are looking at a stained glass window. Uzma says to Pete, 'It's amazing how they can create such a window with different-sized congruent shapes.'
Pete replies, 'Yes, you are right.'
How can this statement be correct?

15.2 Tessellations

HOMEWORK 15B

D

1 On squared paper, show how each of these shapes tessellate. You should draw at least six shapes.

a

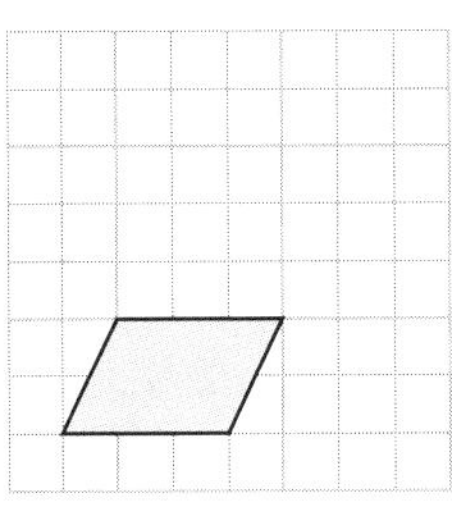

b

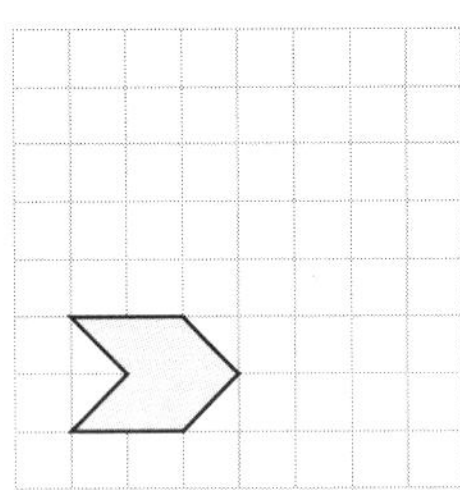

c

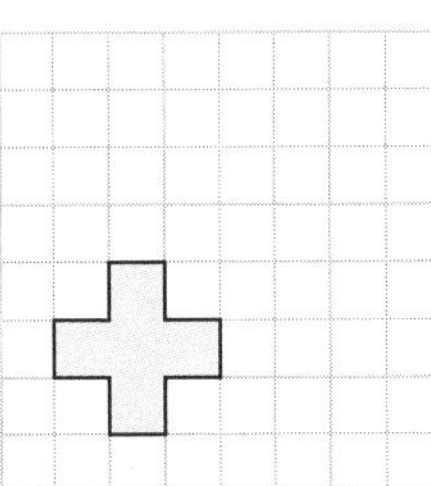

d

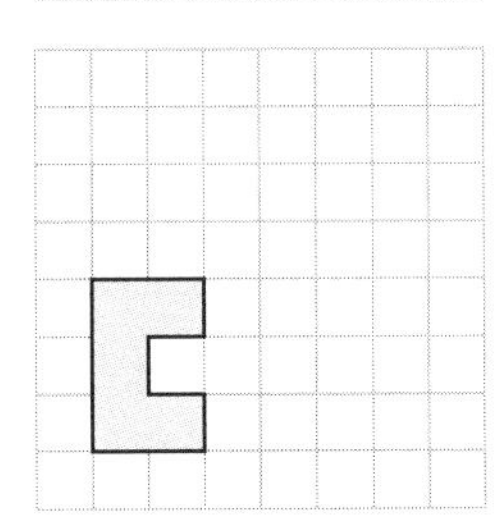

2 Use isometric paper to show how a regular hexagon tessellates.

PS 3 Jeff says, 'All hexagons tessellate.'
Investigate this statement to see if you think it may be true.

AU 4 Explain what is meant by a shape tessellating.

15.3 Translations

HOMEWORK 15C

D

1 Copy each of these shapes onto squared paper and draw its image by using the given translation.

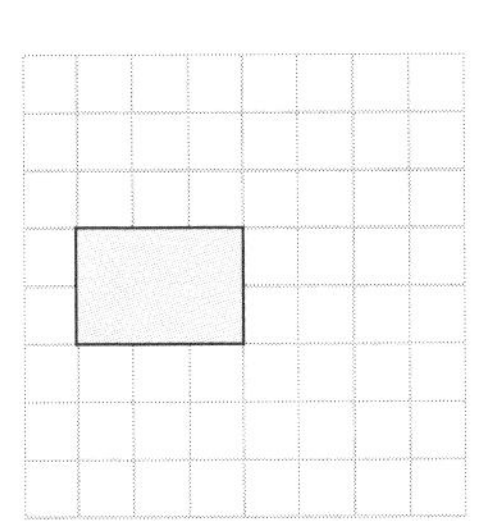

a 4 squares right

b 4 squares up

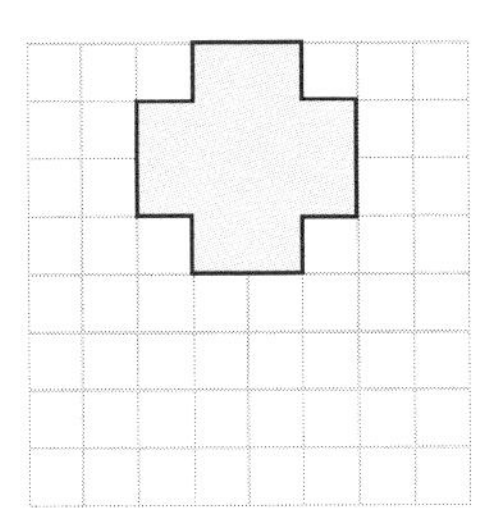

c 4 squares down

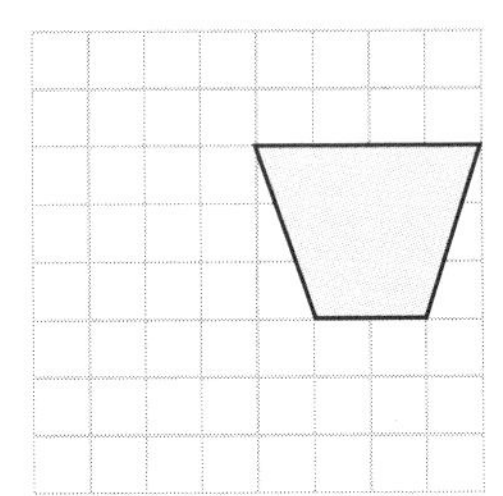

d 4 squares left

2 Copy each of these shapes onto squared paper and draw its image by using the given translation.

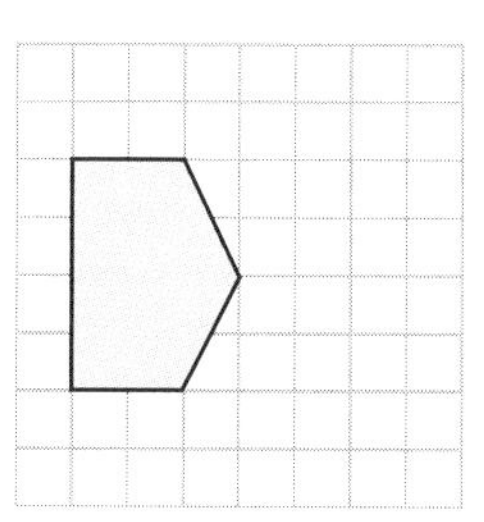

a 3 squares right and 2 squares down

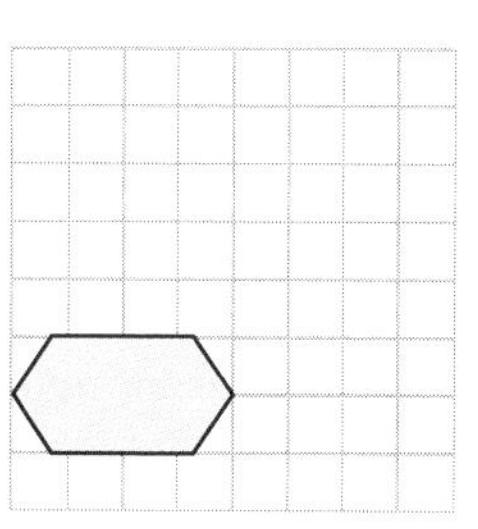

b 3 squares right and 4 squares up

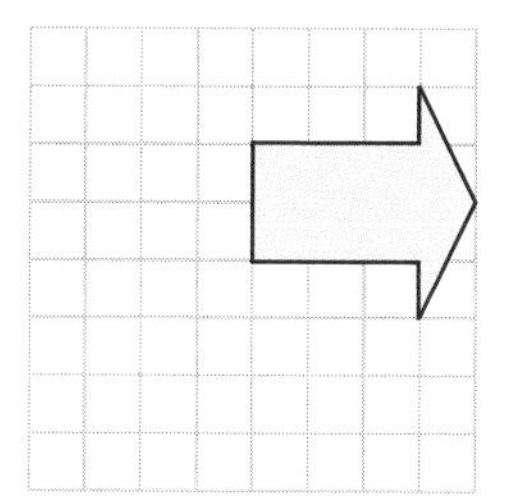

c 3 squares left and 3 squares down

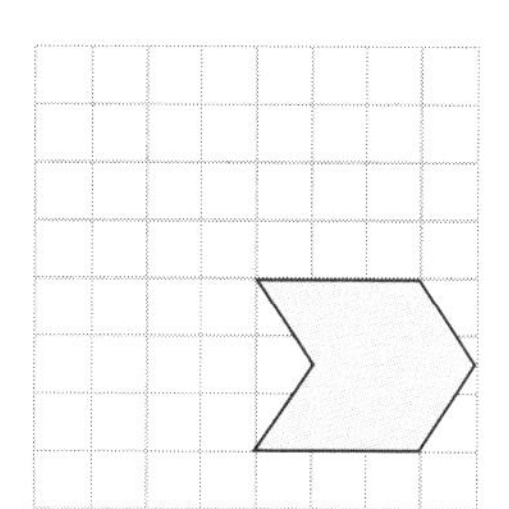

d 4 squares left and 1 square up

3 Describe these translations using vectors.

i A to B **ii** A to C **iii** A to D **iv** B to A **v** B to C **vi** B to D

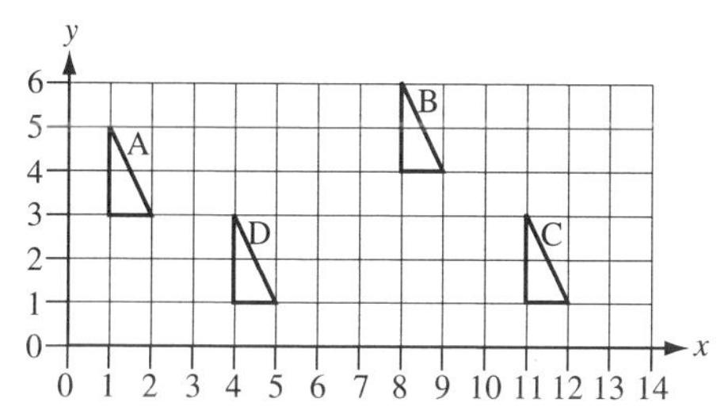

4 **a** On a grid showing values of x and y from 0 to 10, draw the triangle with coordinates A(4, 4), B(5, 7) and C(6, 5).

b Draw the image of ABC after a translation with vector $\binom{3}{2}$. Label this P.

c Draw the image of ABC after a translation with vector $\binom{4}{-3}$. Label this Q.

d Draw the image of ABC after a translation with vector $\binom{-4}{3}$. Label this R.

e Draw the image of ABC after a translation with vector $\binom{-3}{-2}$. Label this S.

PS 5

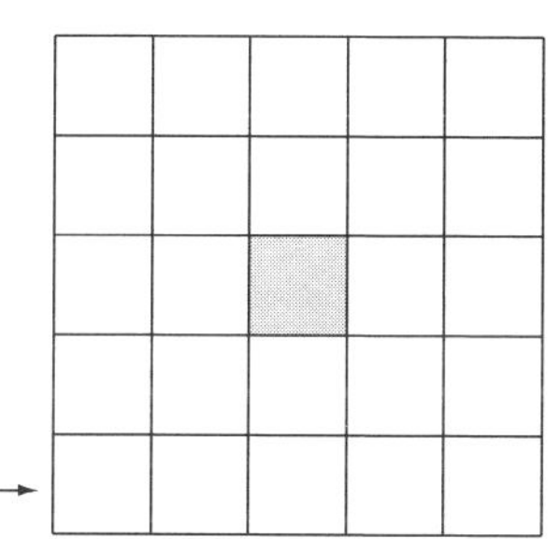

Start / finish →

Write down a series of translations which will take you from the Start / finish around the shaded square without touching it, and back to the Start / finish. Make as few translations as possible.

C

AU **6** A series of translations take you from a starting position back to where you started. If you add together all the numbers shown in the translation vectors, what is the sum?

15.4 Reflections

HOMEWORK 15D

E

1 Copy each shape onto squared paper and draw its image after a reflection in the given mirror line.

a

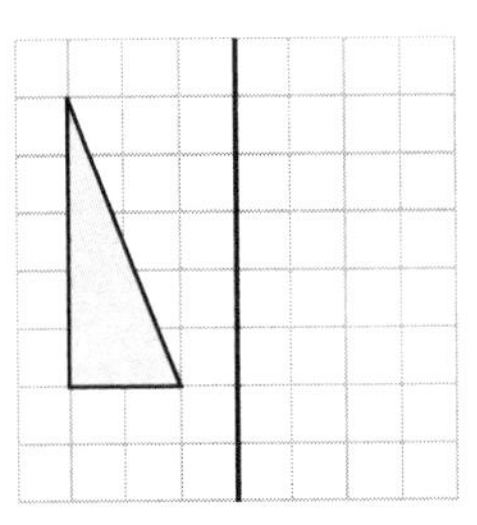

b

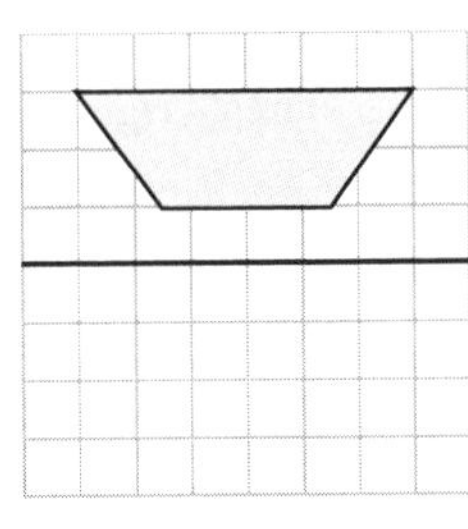

c

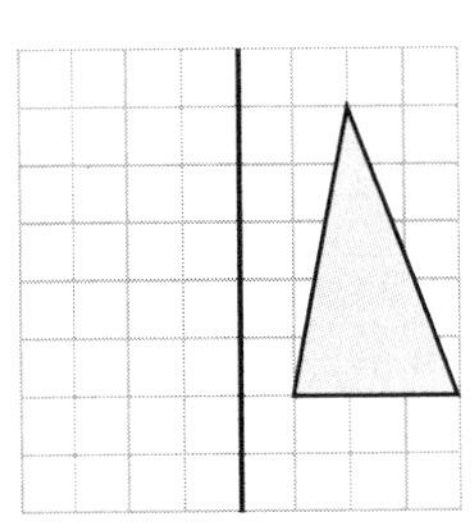

d

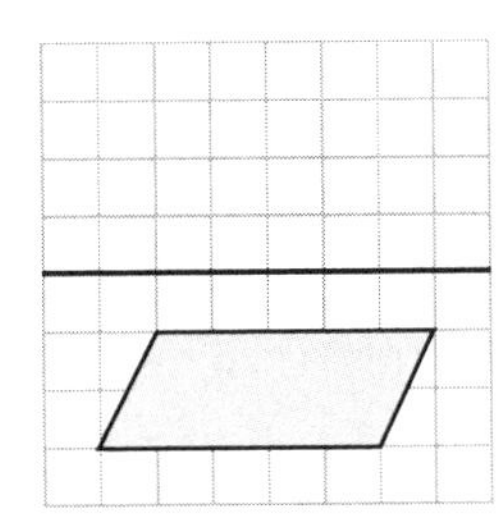

2 Draw each of these figures on squared paper and then draw the reflection of the figure in the mirror line.

a

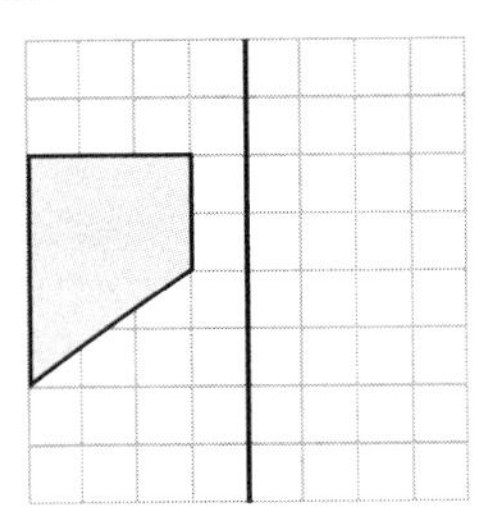

b

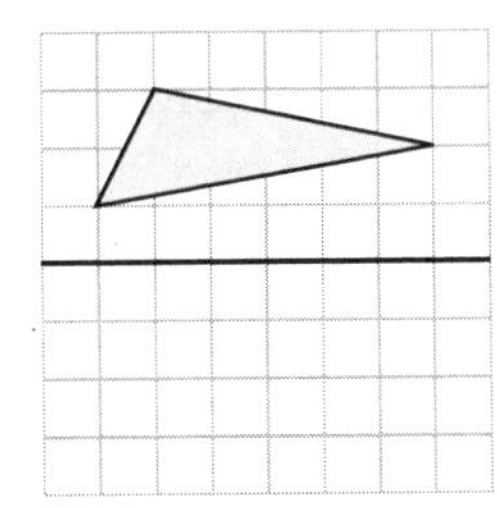

c

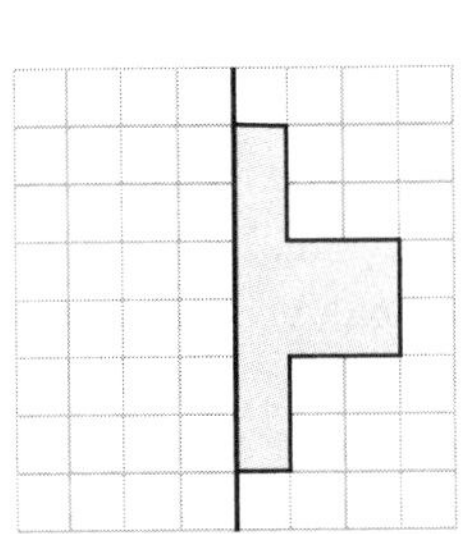

d

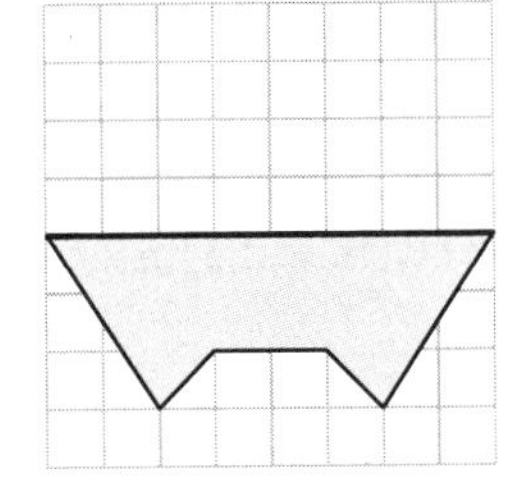

3 Copy this diagram onto squared paper.

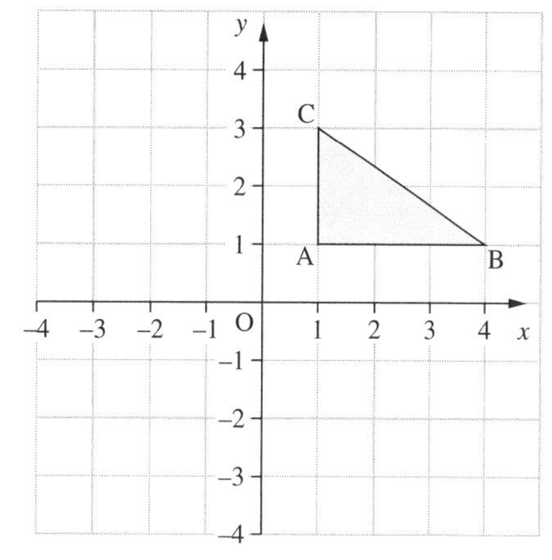

a Reflect the triangle ABC in the x-axis. Label the image R.

b Reflect the triangle ABC in the y-axis. Label the image S.

c What special name is given to figures that are exactly the same shape and size?

PS 4 There are five capital letters that can, when reflected, form another capital letter. Name these letters and what other letter they reflect to form.

AU 5 There is a triangle that, when you draw a reflection on each side, creates the net of a tetrahedron.
What is the name of this triangle?

6 **a** Draw a pair of axes, x-axis from –5 to 5, y-axis from –5 to 5.

b Draw the triangle with coordinates A(2, 2), B(3, 4), C(2, 4).

c Reflect the triangle ABC in the line $y = x$. Label the image P.

d Reflect the triangle P in the line $y = -x$. Label the image Q.

e Reflect triangle Q in the line $y = x$, label it R.

f Describe the reflection that will move triangle ABC to triangle R.

15.5 Rotations

HOMEWORK 15E

1 Copy each of these diagrams onto squared paper. Draw its image using the given rotation about the centre of rotation A. Using tracing paper may help.

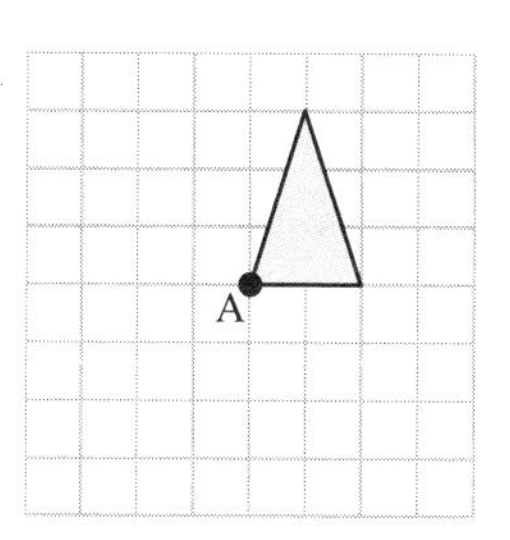

a $\frac{1}{2}$ turn

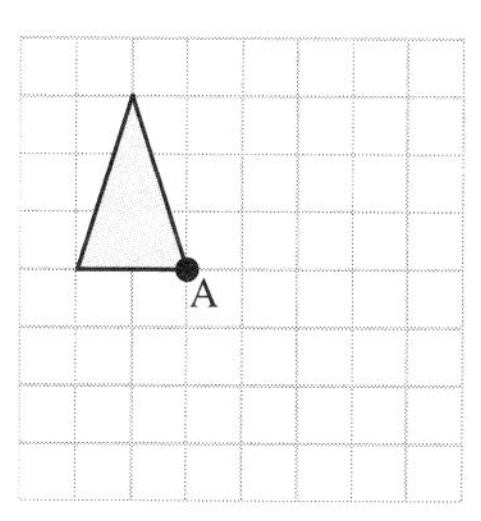

b $\frac{1}{4}$ turn clockwise

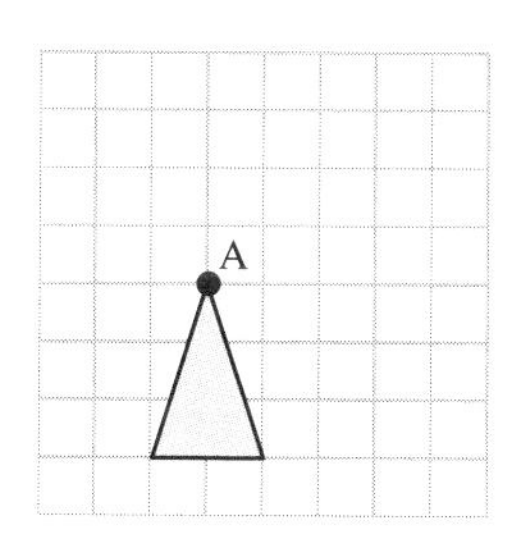

c $\frac{1}{4}$ turn anticlockwise

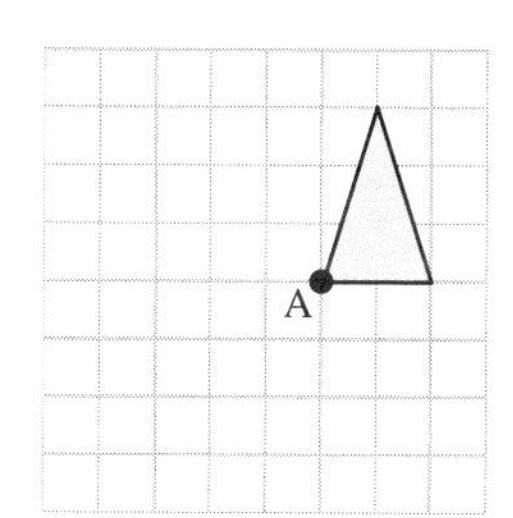

d $\frac{3}{4}$ turn clockwise

2 Copy each of these flags onto squared paper. Draw its image using the given rotation about the centre of rotation A. Using tracing paper may help.

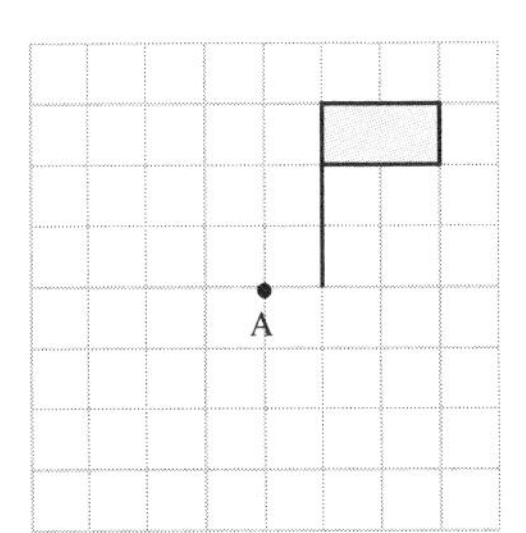

a 180° turn

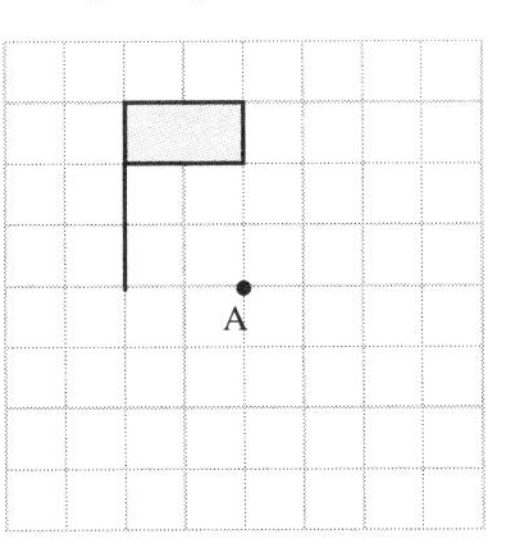

b 90° turn clockwise

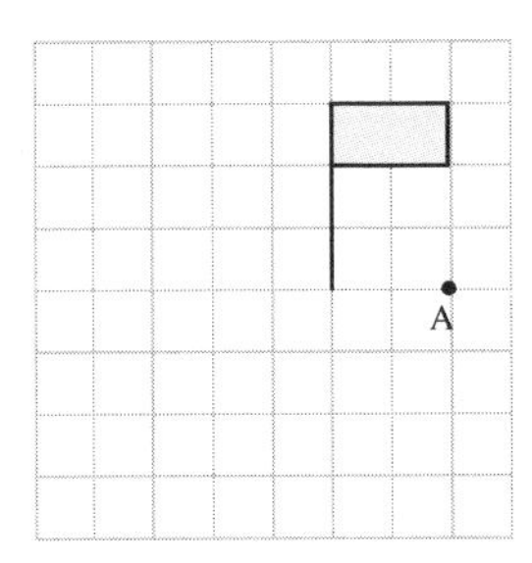

c 90° turn anticlockwise

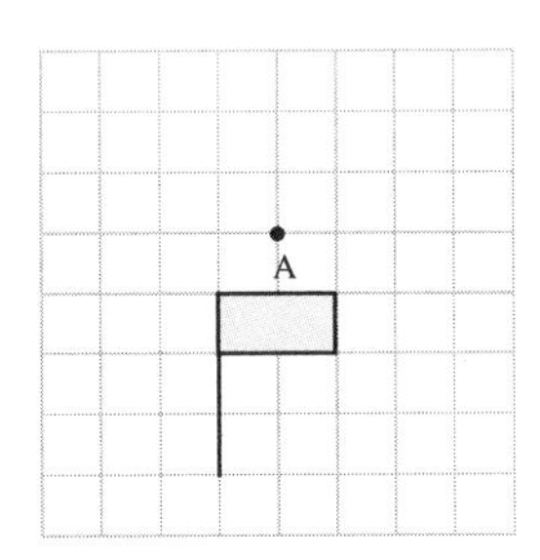

d 270° turn clockwise

3 Copy this T-shape onto squared paper.

a Rotate the shape 90° clockwise about the origin O. Label the image P.

b Rotate the shape 180° about the origin O. Label the image Q.

c Rotate the shape 270° clockwise about the origin O. Label the image R.

d What rotation takes R back to the original shape?

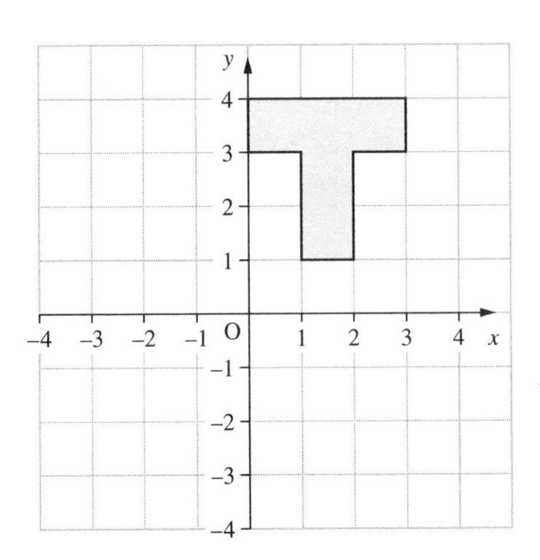

FM 4 A designer came up with the following routine for creating a design:

Start with a rectangle ABCD.
Reflect the rectangle in the line AC.
Rotate the whole shape about the centre point of line AC, clockwise 90 degrees.

From any rectangle of your choice, create a design using the above routine.

PS 5 Choose any one of the triangles below as a starting triangle ABC.

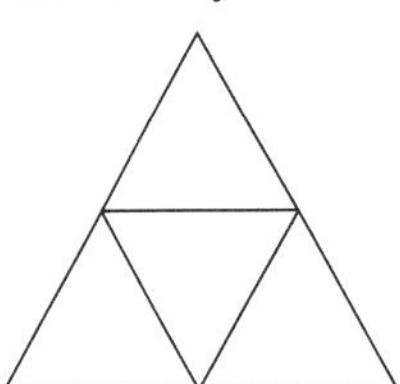

Describe how to keep rotating the shape to get the final diagram shown.

6 Copy the diagram and rotate the given triangle by:

a 90° clockwise about (0, 0)

b 180° about (0, –2)

c 90° anticlockwise about (–1, –1)

d 180° about (0, 0).

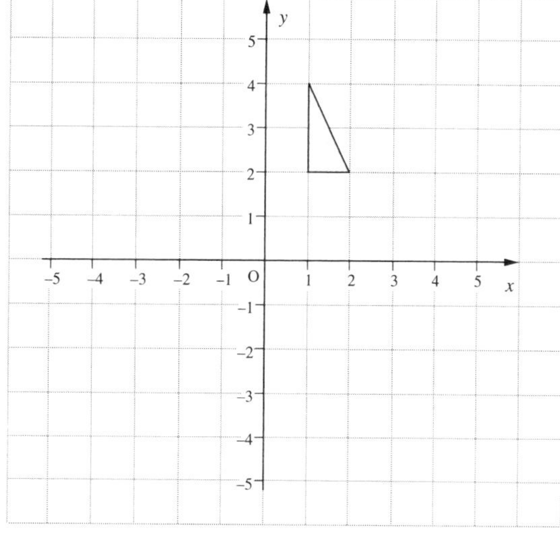

AU 7 Tom said: 'If I rotate a shape, then the image is always congruent.'
Is Tom's statement:

A sometimes true?

B never true?

C always true?

15.6 Enlargements

HOMEWORK 15F

1 Copy each of these figures onto squared paper with its centre of enlargement A. Then enlarge it by the given scale factor using the ray method.

a

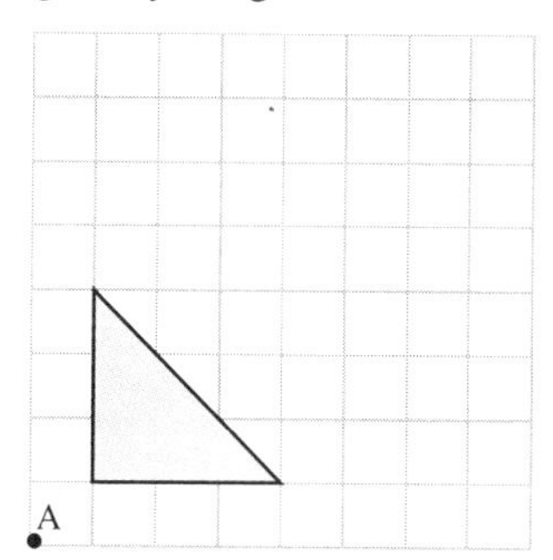

Scale factor 2

b

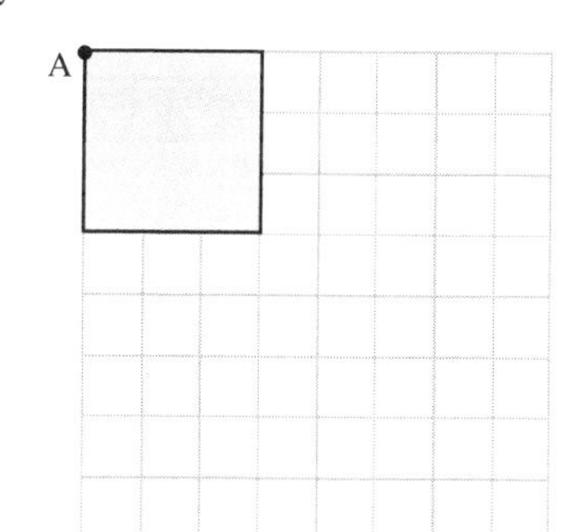

Scale factor 3

2 Copy each of these diagrams onto squared paper and enlarge it by scale factor 2 using the origin as the centre of enlargement.

a

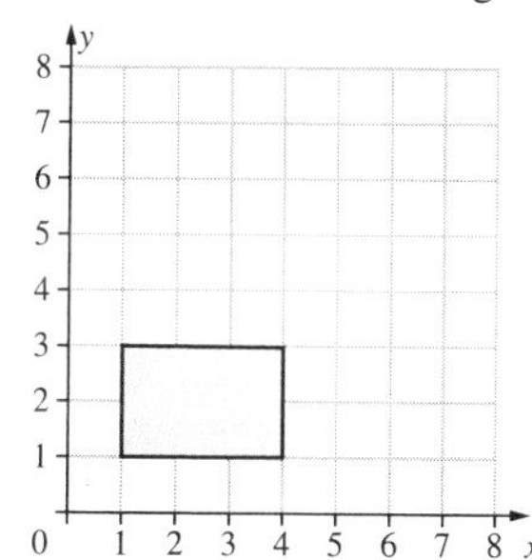

b

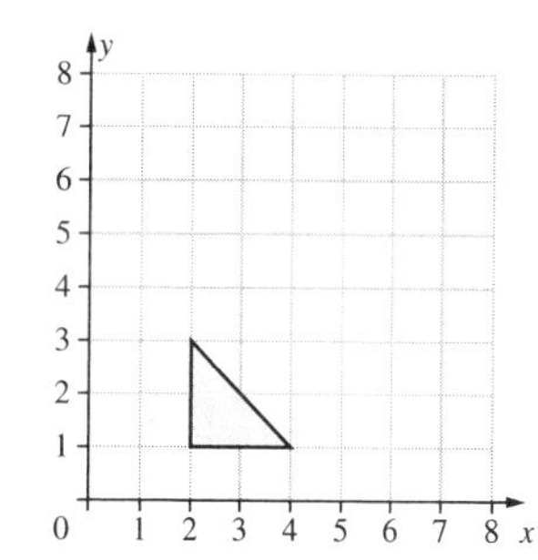

D

c

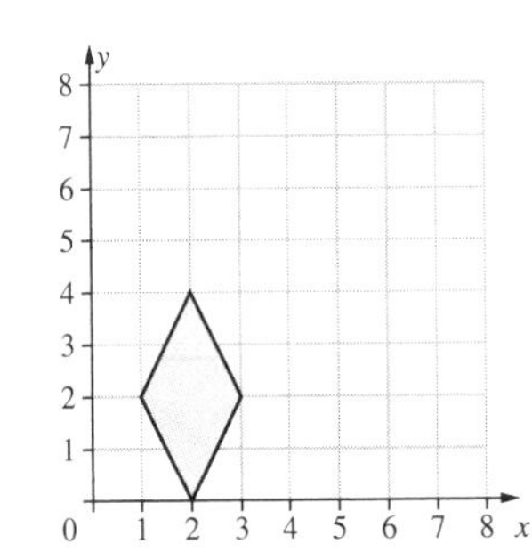

d

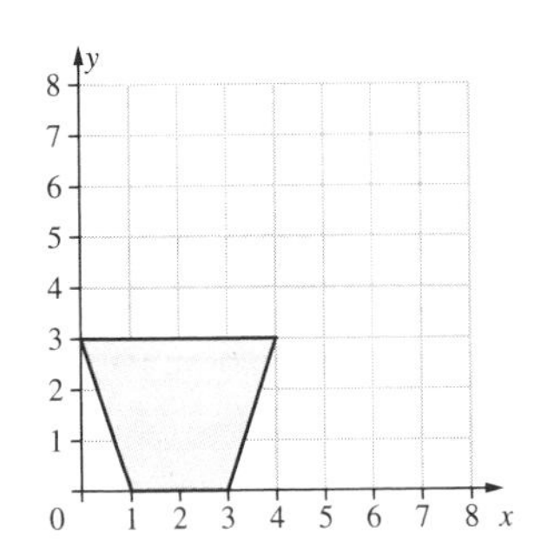

3 Copy each figure below with its centre of enlargement, leaving plenty of space for the enlargement. Then enlarge them by the given scale factor using the counting squares method.

Scale factor 2

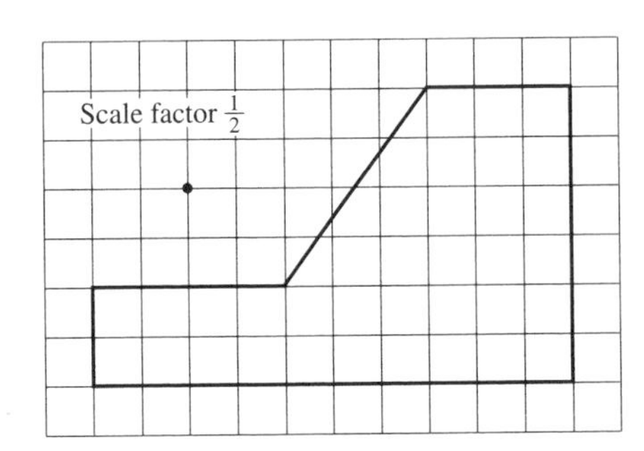

FM 4 A designer is told to use the following routine:

Start with an octagon in the shape of a letter T.

Reflect the T in the small line on the bottom of the T.

Rotate the whole new shape about the midpoint, M, of the small line you have just reflected in.

Enlarge the whole shape by scale factor 2, centre of enlargement point M.

Start with a shape T of your choice and create the design above.

AU 5 Tina enlarged a shape and found the image was congruent to the original. Explain how this might have happened.

C

PS 6 If I enlarge a shape by scale factor 4, by how much will the area of the shape have increased?

Problem-solving Activity

Transformation problem

Look at the diagram showing shape A and shape B.

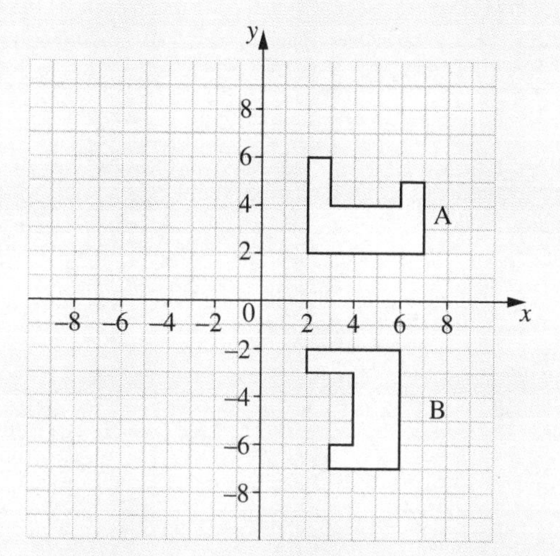

Describe, using three combined transformations, how shape A can be transformed into shape B.

16 Geometry and measures: Constructions

16.1 Constructing triangles

HOMEWORK 16A

D

1 Draw each of the following triangles.

a

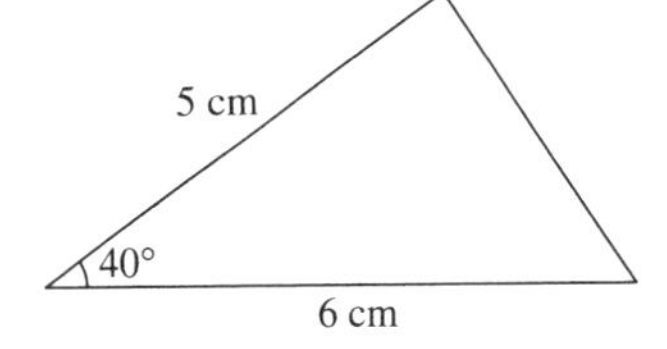

b

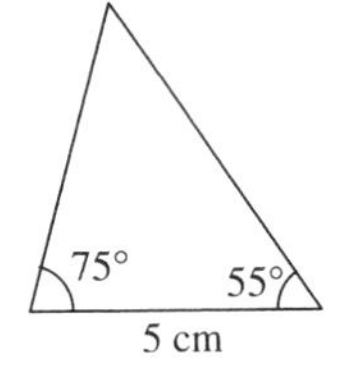

c

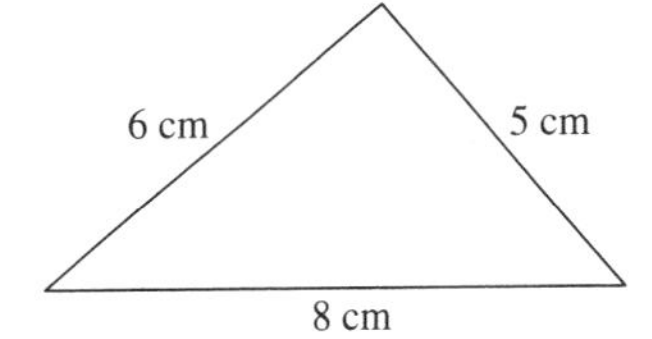

d

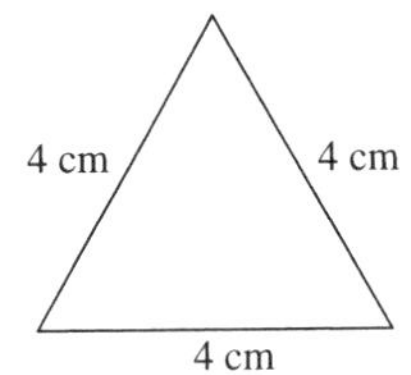

e

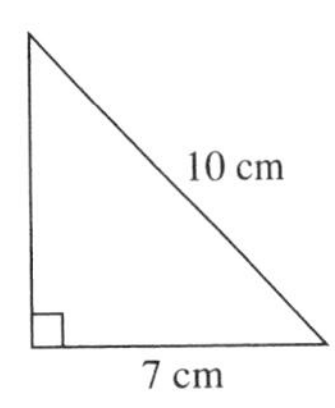

2 Draw a triangle ABC with AB = 6 cm, $\angle A = 60°$ and $\angle B = 50°$.

3

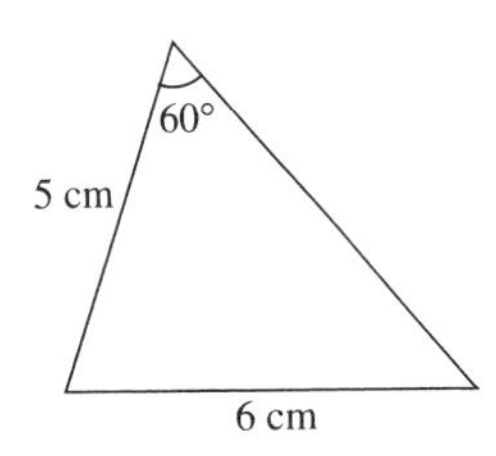

Can you draw this triangle accurately? If so, explain how you would do it.

4 **a** Draw the shape on the right.

b What is the name of the shape you have drawn?

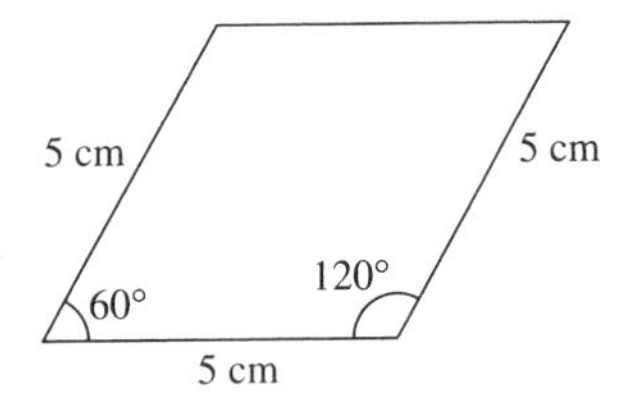

AU 5 Shehab says 'As long as I know two sides of a triangle and one angle I can draw it.' Is Shehab correct? If not, explain why.

PS 6 You are asked to construct a triangle with two sides of 9 cm and 10 cm and an angle of 60°. Sketch all the possible triangles you could construct from this description.

16.2 Bisectors

HOMEWORK 16B

1 Draw a line 8 cm long. Bisect it with a pair of compasses. Check your accuracy by seeing if each half is 4 cm.

2
- **a** Draw any triangle.
- **b** On each side construct the line bisector. All your line bisectors should intersect at the same point.
- **c** See if you can use this point as the centre of a circle that fits perfectly inside the triangle.

3
- **a** Draw a circle with a radius of about 4 cm.
- **b** Draw a quadrilateral such that the vertices (corners) of the quadrilateral are on the circumference of the circle.
- **c** Bisect two of the sides of the quadrilateral. Your bisectors should meet at the centre of the circle.

4
- **a** Draw any angle.
- **b** Construct the angle bisector.
- **c** Check how accurate you have been by measuring each half.

AU 5 The diagram shows a park with two ice-cream sellers A and B. People always go to the ice-cream seller nearest to them. Shade the region of the park from which people go to ice-cream seller B.

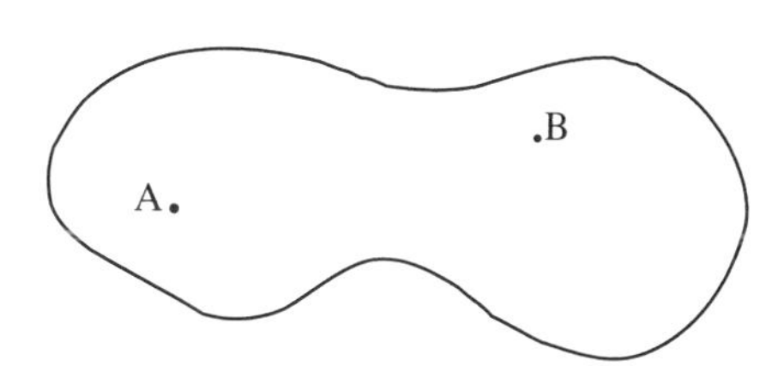

6 Using a straight edge and a pair of compasses only, construct:
- **a** an angle of 15 degrees.
- **b** an angle of 75 degrees.

AU 7 If I construct all the angle bisectors in a triangle, they will meet at a point. Explain why I can draw a circle with this as the centre and why this circle will just touch each side of the triangle.

16.3 Loci

HOMEWORK 16C

1 A is a fixed point. Sketch the locus of the point P when $AP > 3$ cm and $AP < 6$ cm.

2 A and B are two fixed points 4 cm apart. Sketch the locus of the point P for the following situations:
- **a** $AP < BP$
- **b** P is always within 3 cm of A and within 2 cm of B.

3 A fly is tethered by a length of a spider's web that is 1 m long. Describe the locus that the fly can still buzz about in.

4 ABC is an equilateral triangle of side 4 cm. In each of the following loci, the point P moves only inside the triangle. Sketch the locus in each case.
- **a** $AP = BP$
- **b** $AP < BP$
- **c** $CP < 2$ cm
- **d** $CP > 3$ cm and $BP > 3$ cm

C

5 A wheel rolls around the inside of a square. Sketch the locus of the centre of the wheel.

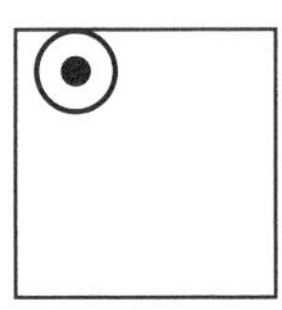

6 The same wheel rolls around the outside of the square. Sketch the locus of the centre of the wheel.

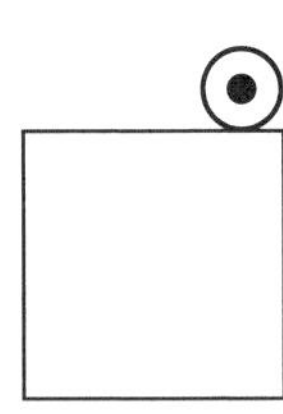

7 Two ships A and B, which are 7 km apart, both hear a distress signal from a fishing boat. The fishing boat is less than 4 km from ship A and is less than 4.5 km from ship B. A helicopter pilot sees that the fishing boat is nearer to ship A than to ship B. Use accurate construction to show the region that contains the fishing boat. Shade this region.

PS 8 On a sheet of plain paper, mark three points A, B and C, about 5 to 7 cm away from one another.
Find the locus of point P where:

a P is always closer to a point A than a point B.
b P is always equal distances from points B and C.

AU 9 Sketch the locus of a point on the rim of a bicycle wheel as it makes three revolutions along a flat road.

FM 10 Each side of a square courtyard is 18 m long. The owner wants to monitor it at night with surveillance cameras around the perimeter. The cameras can be rotated automatically and they have an effective range at night of 10 m. He has asked you how many cameras he needs to buy and where they should be placed. What do you suggest?

HOMEWORK 16D

For Questions 1 to 3, you should start by sketching the picture given in each question on a 6×6 grid, each square of which is 1 cm by 1 cm. The scale for each question is given.

C

1 A goat is tethered by a rope, 10 m long, and a stake that is 2 m from each side of a field. What is the locus of the area that the goat can graze? Use a scale of 1 cm : 2 m.

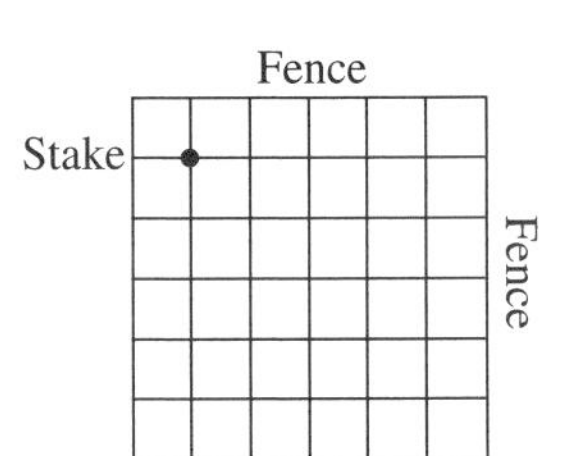

2 A cow is tethered to a rail at the top of a fence 4 m long. The rope is 4 m long. Sketch the area that the cow can graze. Use a scale of 1 cm : 2 cm.

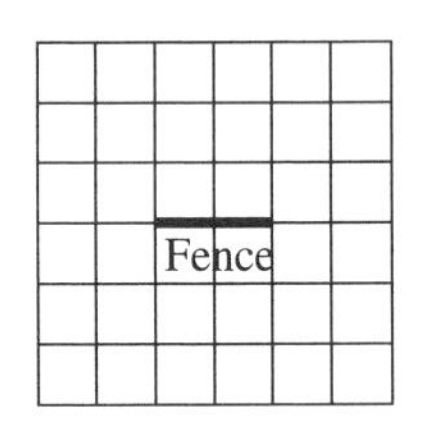

3 A horse is tethered to a corner of a shed, 3 m by 1 m. The rope is 4 m long. Sketch the area that the horse can graze. Use a scale of 1 cm : 1 m.

Tethered here

Shed

Note: For Questions 4 to 6, you should use a copy of the map on page 110. For each question, trace the map and mark those points that are relevant to that question.

FM 4 A radio station broadcasts from Birmingham with a range that is just far enough to reach York. Another radio station broadcasts from Glasgow with a range that is just far enough to reach Newcastle.

a Sketch the area to which each station can broadcast.

b Will the Birmingham station broadcast as far as Norwich?

c Will the two stations interfere with each other?

FM 5 An air traffic control centre is to be built in Newcastle. If it has a range of 200 km, will it cover all the area of Britain north of Sheffield and south of Glasgow?

FM 6 A radio transmitter is to be built so that it is the same distance from Exeter, Norwich and Newcastle.

a Draw the perpendicular bisectors of the lines joining these three places to find where the station is to be built.

b Birmingham has so many radio stations that it cannot have another one within 50 km. Can the transmitter be built?

PS 7 Three radio stations receive a distress call from a boat in the North Sea.
The station at Norwich can tell from the strength of the signal that the boat is within 150 km of the station. The station at Sheffield can tell that the boat is between 100 and 150 km from Sheffield.
If these two reports are correct, then how far away from the helicopter station at Newcastle might the boat be?

AU 8 The locus of a point is described as:

- 5 cm away from point A
- equidistant from both points A and B.

Which of the following could be true?

a The locus is an arc.

b The locus is just two points.

c The locus is a straight line.

d The locus is none of these.

C

C

Glasgow
Newcastle upon Tyne
North Sea
York
Leeds
Irish Sea
Manchester
Sheffield
Norwich
Birmingham
London
Bristol
Exeter
0
50
100
150
200
250 miles
0
50
100
150
200
250
300
350 km

Functional Maths Activity

Loci

1 Put a matchbox and a 2p coin on a table.
Hold the matchbox still and slide the coin around it, keeping the edge of the coin touching the box.
What is the locus of the centre of the coin? Draw a diagram to show it.

> **HINTS AND TIPS**
> Look carefully at what happens at the corners of the matchbox.

2 Now take the tray out of the matchbox, remove the matches and place the coin in the tray.
Slide the coin around in the tray. It can go anywhere and does not have to touch the sides of the tray.
What is the locus of the centre of the coin in this case? Illustrate with a diagram.

3 Suppose you use a 1p coin instead of a 2p coin in questions 1 and 2. How would the loci be different?

4 What would the locus of the centre be like in questions 1 and 2 if you used a marble instead of a coin?

5 When the wheel of a bicycle turns, it rotates around a cylindrical axle.
Between the wheel and the axle is a ring of ball bearings.

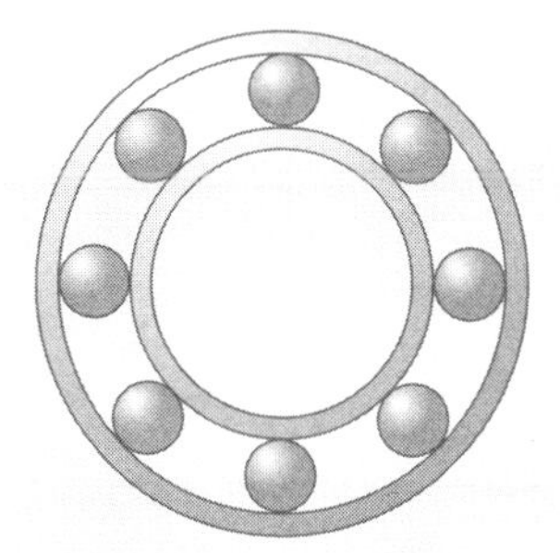

> **HINTS AND TIPS**
> You could model this by sliding a coin (the ball bearing) around a circular jar lid (the axle).

You can see eight ball bearings in this diagram. The axle goes through the centre.
What is the locus of the centre of one of the ball bearings as the wheel turns?

17 Geometry: Pythagoras' theorem

17.1 Pythagoras' theorem

HOMEWORK 17A

For any right-angled triangle:

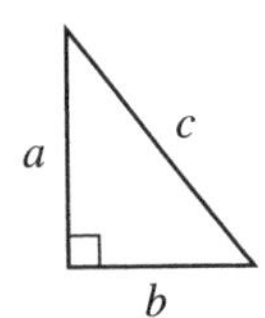

$$a^2 + b^2 = c^2$$

In each of the following right-angled triangles, calculate the length of the hypotenuse, x, giving your answers to one decimal place where necessary.
Note: The triangles in this exercise are not drawn to scale.

1

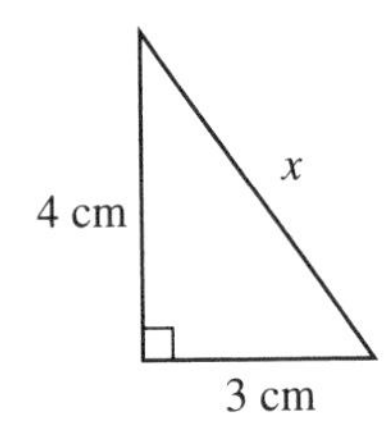

2

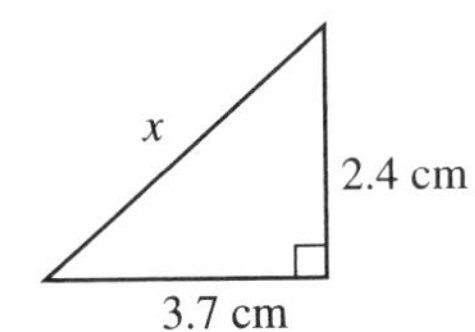

3

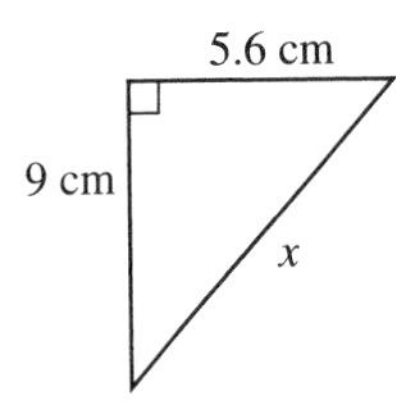

4

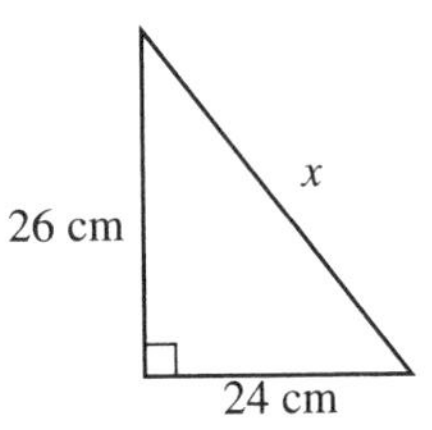

5 Which of the following are right-angled triangles?

a

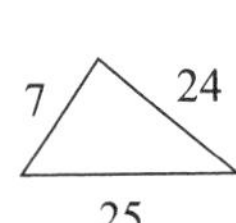

b

15
8
17

c

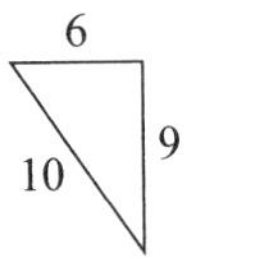

d

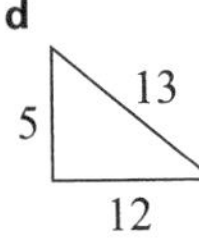

e

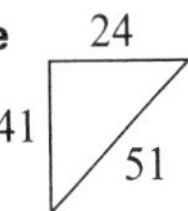

f

10
26
24

g

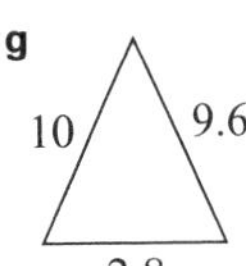

h

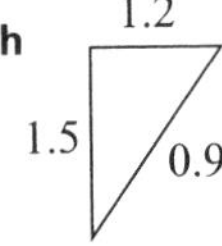

PS 6 The length of the diagonal of a square is 20 cm.
What is the perimeter of the square?

Joe is told that the diagonal of a square is 8 cm.
He immediately says, 'In that case, the area of the triangle must be 32 cm^2.'
Can you explain how he managed to work this out?

17.2 Finding a shorter side

HOMEWORK 17B

For any right-angled triangle:

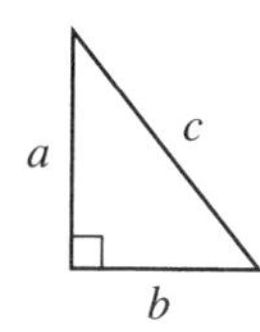

$a^2 = c^2 - b^2$ and $b^2 = c^2 - a^2$

Note: The triangles in this exercise are not drawn to scale.

1 In each of the following right-angled triangles, calculate the length of x, giving your answers to one decimal place where necessary.

a
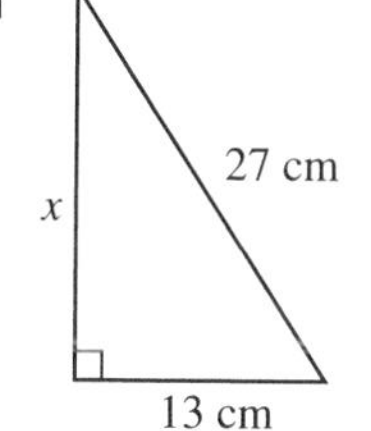

b
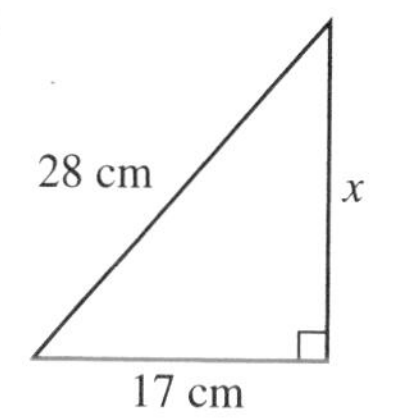

c
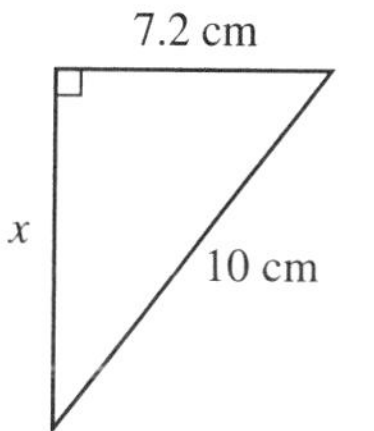

d
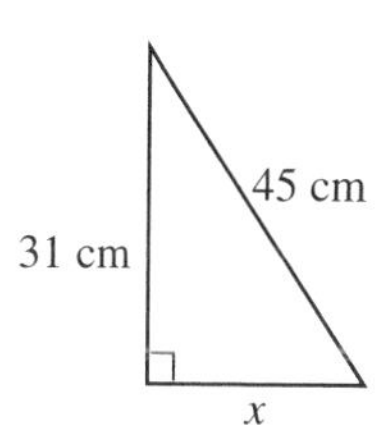

e
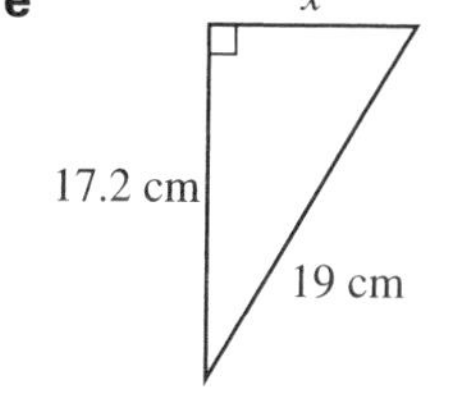

f
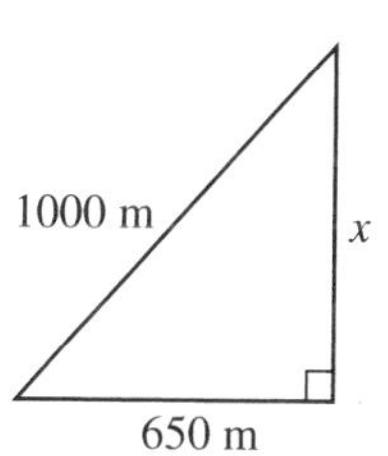

g
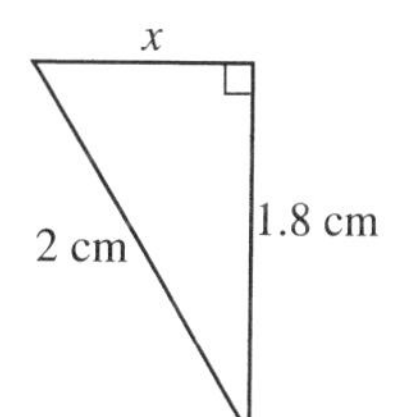

h
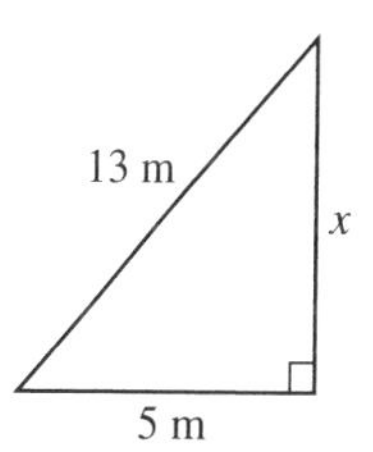

2 In each of the following right-angled triangles, calculate the length of x, giving your answers to one decimal place where necessary.

a
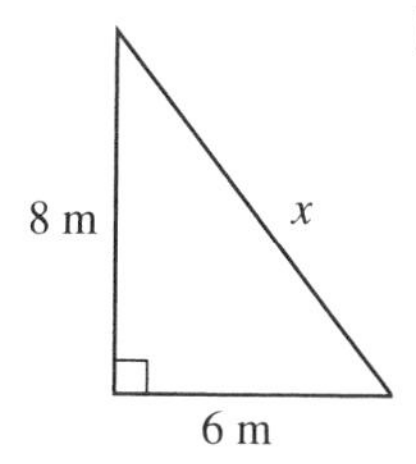

b
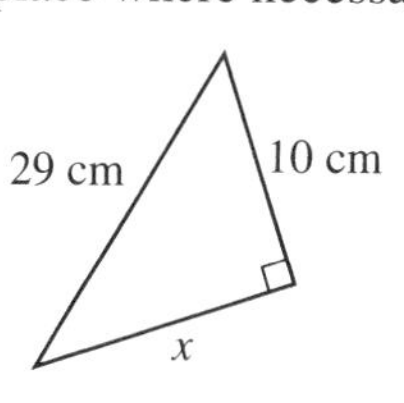

c
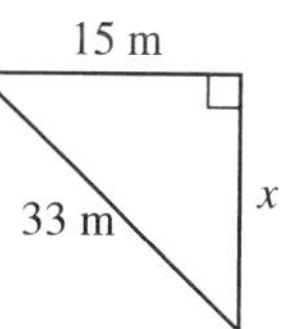

d
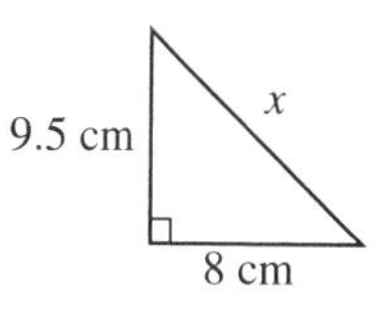

3 The diagram shows the end view of the framework for a sports arena stand. Calculate the length AB.

7 m
16 m
14 m
A
B

C

FM 4 A pilot flies for 300 km and finds himself 200 km north of his original position. How far has he travelled in a horizontal direction?

PS 5 If you draw a semicircle on each side of a right-angled triangle, as shown below, what can you say about their areas?

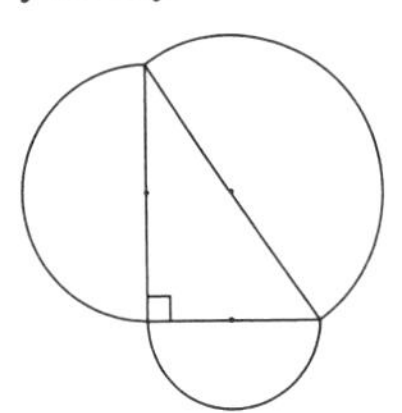

AU 6 Explain how you can tell that the length of CD is 4 cm.

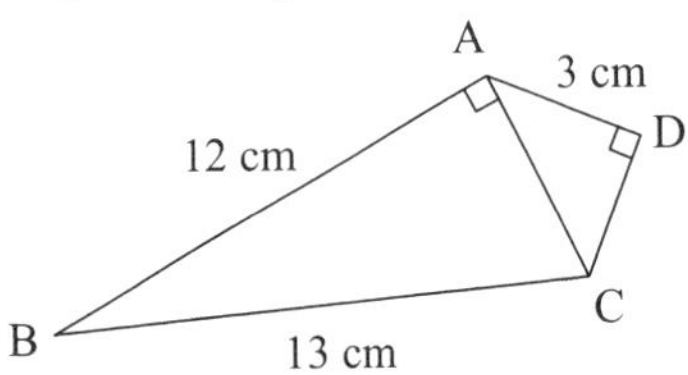

17.3 Solving problems using Pythagoras' theorem

HOMEWORK 17C

C

1 A ladder, 15 metres long, leans against a wall. The ladder reaches 12 metres up the wall. How far away from the foot of the wall is the foot of the ladder? Give your answer to one decimal place.

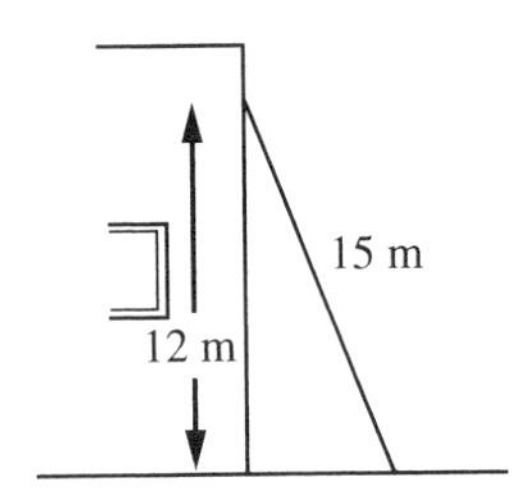

2 A rectangle is 3 metres long and 1.2 m wide. How long is the diagonal? Give your answer to one decimal place.

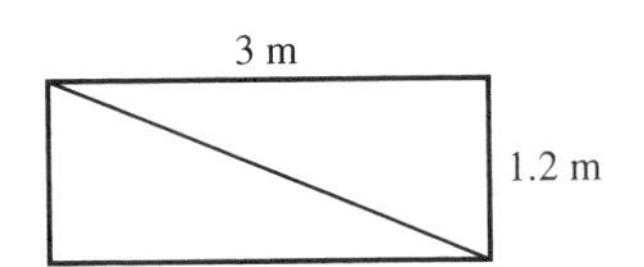

3 How long is the diagonal of a square with a side of 10 metres? Give your answer to one decimal place.

4 A ship going from a port to a lighthouse steams 8 km east and 6 km north. How far is the lighthouse from the port?

FM 5 At the moment, three towns, A, B and C, are joined by two roads, as in the diagram. The council wants to make a road which runs directly from A to C. How much distance will the new road save? Give your answer to one decimal place.

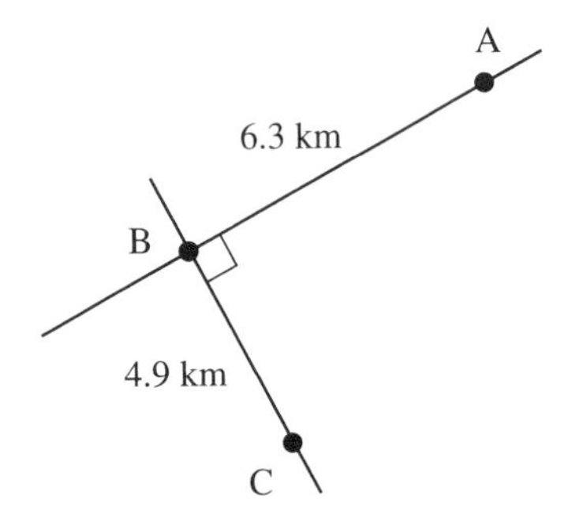

C

6 An 8-metre ladder is put up against a wall.

a How far up the wall will it reach when the foot of the ladder is 1 m away from the wall? Give your answer to one decimal place.

b When it reaches 7 m up the wall, how far is the foot of the ladder away from the wall? Give your answer to one decimal place.

7 How long is the line that joins the two coordinates A(1, 3) and B(2, 2)? Give your answer to one decimal place.

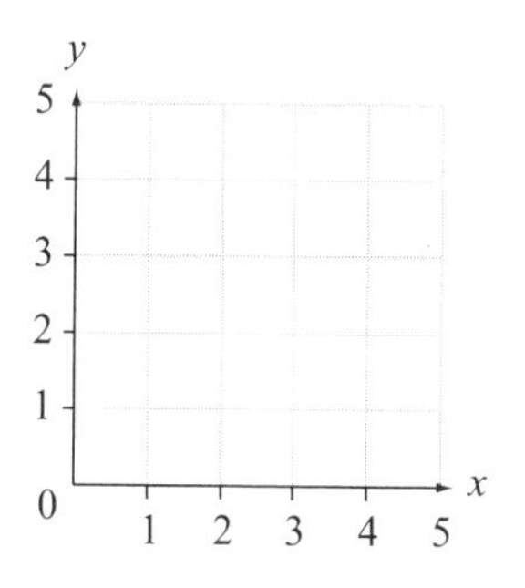

8 A rectangle is 4 cm long. The length of its diagonal is 5 cm. What is the area of the rectangle? Give your answer to one decimal place.

9 Is a triangle with sides 9 cm, 40 cm and 41 cm a right-angled triangle?

10 How long is the line that joins the two coordinates A(–3, –7), and B(4, 6)?

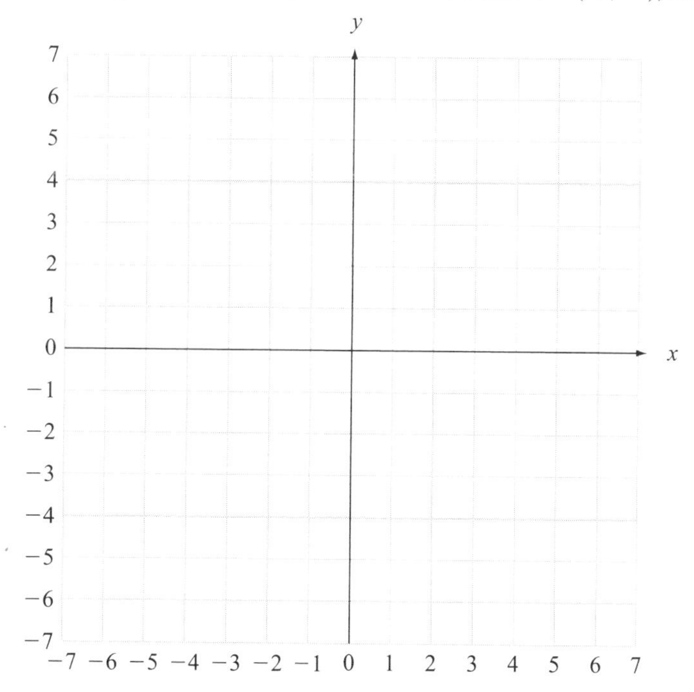

AU 11 A helicopter rises vertically from the ground 200 m, then flies due north for 300 m before turning due east for 500 m.
How far is the helicopter now from its starting point?

PS 12 A 13-cm pencil fits exactly diagonally in a cylinder.
I know that the dimensions of this cylinder are integer values.
What are the two possible dimensions of this cylinder?

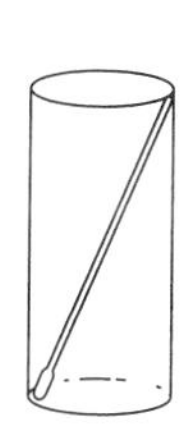

Problem-solving Activity

Pythagoras' theorem

An interesting spiral shape can be constructed with right-angled triangles, as shown in the diagram below.

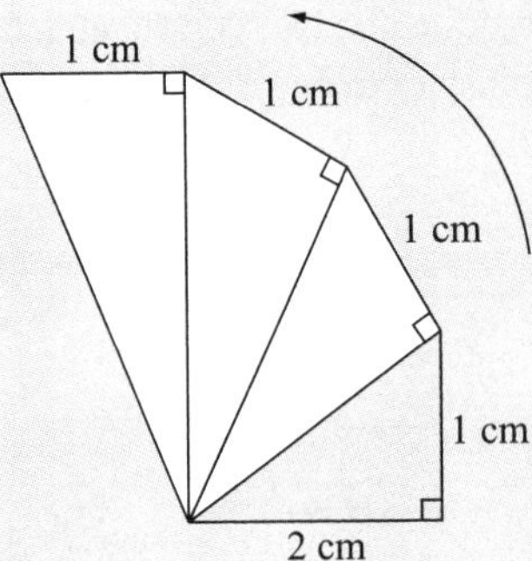

To make this shape, start with a right-angled triangle of base 2 cm and height 1 cm (the shaded triangle at the bottom of the diagram).

The second right-angled triangle is built on top of the hypotenuse of the first triangle, and has a height of 1 cm.

The shape grows by continuing to put right-angled triangles (each with a height of 1 cm) on top, as shown in the diagram.

1 Draw the shape as far as you can go.
2 Measure the hypotenuse of the last triangle to be drawn.
3 Now calculate what that length should have been.

Now look at the following shape, which is built up from a first right-angled triangle which has a hypotenuse of 5 cm and height 1 cm. This time, each triangle added underneath is built under the base of the previous one, with the previous base length becoming the new hypotenuse.

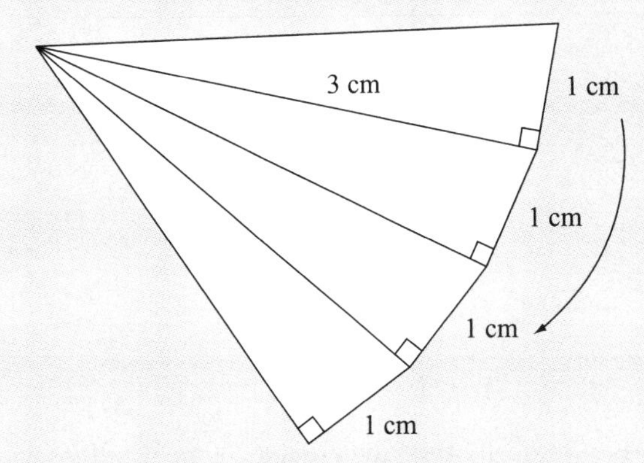

4 Continue drawing the shape as far as you think you can.
5 Measure the length of the smallest base that you end up with.
6 Calculate the length of the final base.

William Collins' dream of knowledge for all began with the publication of his first book in 1819. A self-educated mill worker, he not only enriched millions of lives, but also founded a flourishing publishing house. Today, staying true to this spirit, Collins books are packed with inspiration, innovation and practical expertise. They place you at the centre of a world of possibility and give you exactly what you need to explore it.

Collins. Freedom to teach.

Published by Collins
An imprint of HarperCollins*Publishers*
77–85 Fulham Palace Road
Hammersmith
London
W6 8JB

Browse the complete Collins catalogue at
www.collinseducation.com

10 9 8 7 6 5 4 3 2 1

ISBN-13 978-0-00-733991-4

British Library Cataloguing in Publication Data
A Catalogue record for this publication is available from the British Library

Commissioned by Katie Sergeant
Project managed by Patricia Briggs
Edited by Brian Asbury
Cover design by Angela English
Concept design by Nigel Jordan
Illustrations by Wearset Publishing Services
Typesetting by Wearset Publishing Services
Production by Simon Moore
Printed and bound by L.E.G.O. S.p.A. Italy

Acknowledgement
With thanks to Chris Pearce (Teaching and Learning Advisor, North Somerset).

Important information about the Student Book CD-ROM
The accompanying CD-ROM is for home use only. You cannot copy or save the files to your hard drive and it will work only when placed in the CD-ROM drive.